CORPORATE GREENING 2.0: CREATE AND COMMUNICATE YOUR COMPANY'S CLIMATE CHANGE AND SUSTAINABILITY STRATEGIES

E. BRUCE HARRISON

Author of the Landmark Business Communications Book
Going Green

PublishingWorks, Inc.
2008

Copyright © 2008 by E. Bruce Harrison

All rights reserved. No part of this book may be reproduced or transmitted in any form or by any means, electronic or mechanical, including photocopying, recording, or by an information storage and retrieval system—except by a reviewer who may quote brief passages in a review to be printed in a magazine or newspaper—without permission in writing from the publisher.

Second Printing, 2008

PublishingWorks, Inc.
60 Winter Street
Exeter, NH 03833
603-778-9883
For Sales and Orders:
1-800-738-6603 or 603-772-7200

Designed by Barry Spector and Kat Mack
Cover design by Barry Spector and Jessica Tillyer

LCCN: 2008920181
ISBN: 1-933002-70-0
ISBN-13: 978-1-933002-70-5

Printed in Canada.

CORPORATE GREENING 2.0: CREATE AND COMMUNICATE YOUR COMPANY'S CLIMATE CHANGE AND SUSTAINABILITY STRATEGIES

Acknowledgments

My original ambition, to be a newspaper reporter, has never left me, and this book—like my earlier efforts, especially *Going Green*—is the end product of the process of writing for deadlines and thinking in between. I acknowledge with gratitude those who have both challenged and have given me the opportunity to think alongside them and to write with their encouragement. Special thanks go to my associates in EnviroComm, and for many years before in the company that bore my name; to my clients in companies with whom I have worked and from whom I have learned for so many years. The accent for me is always on the learning. My personal working mantra is to listen, to learn, and to leverage what I learn into something useful. That is certainly the case since 1998 when teamwork on management issues at Navistar, with primary focus on the sustainable role of diesel engine and vehicle technology—resulting in the *Green Diesel Technology* program—provided me with a great opportunity to learn, to contribute and to share in the joy of discovered solutions and meaningful success. My professional belief is that corporate success happens in the nexus of stakeholder approval and top management vision. I would single out in this regard the two outstanding chief executives who led the company during these years—John Horne and Dan Ustian—whose vision and steady guidance inspired all of us on the special project of assuring that diesel engines, trucks and school buses would contribute lasting value as reliable, energy-efficient, and low-emitting public and private transportation; and, while I run the risk of neglecting to convey my respect and appreciation to all with whom I am and have been associated with Navistar, I must mention with deep thanks Brian Whalen, Greg Elliott, Patrick Charbonneau and Dr. William Bunn for their encouragement and friendship, and for her initial trust in my ability to collaborate with her and Navistar, the person who introduced me to the challenges faced by diesel a decade ago, Maril MacDonald, who headed corporate communications at the company before launching her current, successful consultancy in Chicago. In order to provide some baseline orientations, some glimpses of past situations relevant to the perspective of this book, I have included in the volume a few of my previous commentaries as they appeared in various business, public relations and Internet publications and outlets. I want to thank my journalism colleagues, friends and editors—starting with Paul Holmes who picked up my voluble views way back when; Julia Hood, serving as the perceptive editor at *PRWeek* in recent years; Kevin McCauley of the O'Dwyer's group, whose encouragement has helped me to focus on interpreting sustainable communications for the public relations community during the rapid rise of this topic since 2004—and my mentors and fellow travelers, notable among whom is George Carpenter, the father of corporate sustainability, making the business case at Procter & Gamble as early as 1990 and making the United Nations Rio scene with me in 1992, as well as Terry Yosie and Ernie Rosenberg, who have moved along the sustainability trail in an impressive manner after our collaborations in counseling a decade ago, and whose thinking and ability to weave wonderful outcomes from intimidating circumstances inspire me substantially.

No man is an island and no book is launched without a lot of patience, guidance and help to the author from friends, professionals and family members. My thanks go to Barry Spector, Jessica Tillyer, Jeremy Townsend, Ken Nasshan, Phil and Sam Harrison, Diana Roberts, Jim Sloan and many others who, one way or another, got me to the publication shore. I was especially lucky to have at my side during the writing and production of this book, overcoming problems, reading, researching, expediting to meet deadlines and catching mistakes (including a huge mistake in the title I had originally picked for the book) Claudia Spain-Grove. Without Claudia's consistently smart help, I'd still be fiddling with the manuscript.

Bruce Harrison

Corporate Greening 2.0 is dedicated to Patricia and to our family.

In memory of John Phillip Harrison

Contents

iv Acknowledgments

1 PREFACE
Strategic Notes on Corporate Greening in American C-Suites

17 CHAPTER ONE
Dealing With the Comeback Kid: Five Factors in Play as Executives Zero in on Climate Change

29 CHAPTER TWO
How Business Arrived Here: The Hot and Cold Trail of Sociopolitical Greening

43 CHAPTER THREE
Is Your Company Still Going Green? Dealing with the Signs on the Road to Sustainability

51 CHAPTER FOUR
Corporate Sustainability: Balancing the Tripartite Accountabilities

59 CHAPTER FIVE
Seat at the C-Suite Table: Putting Climate Change into Competitive Business Strategies

71 CHAPTER SIX
Changing Orientation: Sustainable Course for Corporate Communicators

83 CHAPTER SEVEN
Climate Change Raises Pressure: Investors' New Focus on Sustainability

97 CHAPTER EIGHT
Where They Go to Learn About Your Sustainability

105 CHAPTER NINE
Carbonomics: Adjusting the Corporate Strategy

123 CHAPTER TEN
Communicate Your Carbon War Strategies: Companies Post their Climate Change Positions

137 CHAPTER ELEVEN
Communication Comeback: From Greenwashing Defense to Collaborative Offense

153 CHAPTER TWELVE
Corporate Greening 2.0: Green Walking into the Future

165 CHAPTER THIRTEEN
The Bridge to Sustainability: What to Expect, What's Relevant

177 GUIDE TO CORPORATE CLIMATE CHANGE AND SUSTAINABILITY POSITIONS

221 APPENDIX
Climate Change Advice to Policy Makers
Dingell's Tax Proposal
Terra Choice's "Six Sins of Greenwashing"
The Page Principles

231 INDEX

"For us, as a company, the scientific debate about climate change is over. The debate now is about what can we do about it. Businesses, like ours, should turn CO_2 management into a business opportunity and lead the search for responsible ways to manage CO_2, use energy more efficiently and provide the extra energy the world needs …"

—*Jeroen van der Veer, CEO, Shell*[1]

"Most scientists believe that greenhouse gas emissions from human activities are influencing the Earth's climate. Although there's much to learn about the cause and effect of climate change, consensus is building that steps should be taken now to reduce those emissions. Duke Energy shares that view …"

—*Company Web site, 2007*

---

[1] Quote from the chief executive's statement Duke Energy Web site, accessed March 21, 2008.

Preface

Strategic Notes on Corporate Greening American C-Suites

Three American events in the first decade of the 21st century turned a lengthy discussion about climate change into an all-out war on carbon. A catastrophic storm ended a science debate. National elections put green activists back in charge of Congress. And the communications skill of an ex-Vice President of the United States won public attention, a Hollywood Oscar and a Nobel Prize. By 2007, with hurricane Katrina providing the push for scientific rationales, a Democratic Congress taking up federal legislative reins, and Al Gore providing the arousing message of global warming peril, a warring drive to constrain greenhouse gas—led by carbon dioxide—emissions was well under way. Congress began enacting laws to limit sources of emissions. Consensus formed around a law aimed at automobiles made in America, incentives to encourage lower-impact technology, energy use and fuel sources, and new regulatory structures to cap, trade, sequester and otherwise control the flow of carbon emissions.

The unfolding carbon war story now involves American business reactions and interactions. It is about how companies get involved in sociopolitical issues, how they become allies of one another in a common cause, how they relate to their respective stakeholders, with government at many levels and with those who in previous and different circumstances have been adversaries.

It is the story of business awakening to reality. A *Business Week* cover article caught the essence in General Motors' corporate greening epiphany. At an auto show speech in 2007, CEO Richard Wagoner surprised the public with a vow to develop a new electric car that would leapfrog competitors and reach customers in three years. "After years of avoiding the future," the *BW* article said, "[Wagoner] understood oil prices were not going to return to earth, global warming was a de facto political reality, and Washington was serious about imposing tougher fuel economy rules on his industry. GM would have to live green or die."[2]

In short, the business-side story and the focus of *Corporate Greening 2.0* is how C-suite executives have begun adjusting to uncommon sociopolitical change: forming, managing and communicating their climate change and corporate sustainability commitments and strategies.

This is an intriguing story as well as familiar territory for me, plugged into my counseling on environmental, energy and sustainability issues at the corporate level. My interest in "going green" which is now "going sustainable" covers many years, beginning with work with chemical industry executives in response to Rachel Carson's *Silent Spring*.[3] That was in 1962. My job as vice president/public affairs at Freeport Sulphur Company in New York (now Freeport-McMoran in New Orleans) subsequently kept environmental issues on my plate both in Washington DC and in our mining and chemical operations. That was followed by a counseling career focused on environmental and energy issues, working inside more than 50 corporations with senior environmental and engineering people as well as with CEOs and public relations officers. By founding EnviroComm International in 1992, I was able to expand my work and that of my very capable, green-specialized team to Europe, South America, Asia and Scandinavia. A somewhat rigorous learning experience, working out strategies and solutions alongside the finest minds in the business community, continues and gives me a real-world, reliably informed perspective to approach climate change and sustainability communications.

Much has happened since I went with a business delegation to the Rio Earth summit in 1992, came back and wrote *Going Green*. Environmental and energy management is operational in most companies. Green issues remain but progress is quantifiable. Despite any rhetoric to the contrary, the American business community has over the past two decades led the world in innovations that deal with green challenges in an advanced economy. In 2005, on a 14-day corporate environmental group visit to

---

[2] *Business Week*, May 15, 2008, "GM: Live Green or Die," by David Welch. Also accessible at http://www.businessweek.com.

[3] *Silent Spring* was introduced as a series of articles in *The New Yorker* magazine before publication by Houghton Mifflin Company in 1962. It is interesting to note that, a quarter century before the all-out war on carbon and global warming, Ms. Carson's focus was on carbon, the organic atoms "linked with the basic chemistry of all life, they lend themselves to the modifications which make them agents of death." Insecticides, especially DDT, a chlorinated hydrocarbon, were the major target of the writer's attack on industrial products and their impact on wildlife and humans.

Preface
Strategic Notes on Corporate Greening in American C-Suites

China, where we met with government, university and business leaders, I understood very clearly how far nations with growing economies and unresolved environmental and energy challenges will have to go to get to the place American corporations are now in managing such challenges.

Now, the involvement of American business in the sociopolitical, socioeconomic test of climate change is under way. If the past is any guide, our nation's business executives and senior level communicators will influence favorable, win-win outcomes, handling challenges to society and to business in ways that will instruct their counterparts as well as policy makers in other countries. The corporate sustainability goal will be economic and social compatibility. The strategies must be worked out within each company, mindful of respective markets, stakeholders and business plans. All this has stimulated my rethinking of green accountability, and resulted in this book. I hope it is useful to C-suite thinkers, and particularly those who are responsible for stakeholder engagement and communications at the higher strategic levels. Feel free to scan and find what's relevant to your situation. And if you don't have time to read the book, here are my CliffsNotes:

The climate change science debate has been overtaken by events that business cannot ignore. If companies could just play by sound-science rules, it could make decisions differently than when forced to play by government—or sociopolitical—rules. There remains considerable resistance and skepticism about global warming in and outside the business community.[4] But, as Wagoner recognized at GM, companies must deal with the reality that legislation is on the table, under discussion and being enacted. Rules regulating business are well along the impact track. Polls show people want business to get with the climate change program now. Stakeholder opinion, including company investor opinion, has tipped into activism. Events—weather disasters, the Gore recognitions, 2006 congressional elections—have blown aside the input of science, like it or not. Companies affected have little choice but to deal

---

[4] For an impressive skeptical presentation, you may want to look at the documentary from the UK entitled "The Great Global Warming Swindle"—accessible (3/1/08) at http://video.google.com/videoplay?docid=449956022478442170 or Google the title to get it. Among the better organized think-tank programs, The Heartland Institute has research, papers and an ongoing campaign that aggressively attacks the popular acceptance of the science and is sensitive to the climate-change pressures on business. Heartland has directly challenged former Vice President Al Gore on his representations of global warming evidence and exhortations. See http://www.globalwarmingheartland.org or Google "Heartland Institute" and browse the work of Joe Bast and his associates.

with it—hopefully for better rather than worse. Withdrawal from—or attempting to wait out—the war on carbon is a risky business option that few companies will choose.

The need for a concept such as corporate sustainability derives from the interaction of environmental and energy performance with financial performance. "Sustainability" started in the 1980s as a national policy idea. It meant keeping environmental and economic options open for future generations in all countries. At the 1992 United Nations Earth Summit, in which I participated as part of the American business delegation, sustainable development was the key topic. Nature and environment were the key drivers. Now it's moved nearer the top of corporate agendas and energy has made it the hot topic. Climate change—essentially, public awareness and government action on global warming—has added carbon constraint to management's obligations in pollution control and has changed the energy equation.

Arguably, this is the most potent sociopolitical fusion confronting corporate America. It drives the question from stakeholders: is this business, adjusted for climate change, sustainable? For some companies, the climate-change–carbon-war dynamics set up potentially depleting financial implications; for others, lucrative opportunities. For all companies, the concept of sustainability and the necessary adjustment to the reality of constraining carbon emissions are requiring management to give serious thought to a comprehensive strategy on climate change, to minimize risks and to zero in on ways to beat competitors.

A company faces several levels of risk in the focus on climate change and sustainability. The successful company now has to deal with the prospect of new cost and requirement factors that adversely affect financial performance, implications that are adverse to the company's employee and external stakeholder relationships, and—running through all of these in an important, high-level manner—corporate communications factors that are different from what's gone on before.

On the other side of the coin are potential rewards. That's the business-critical side of the story for any company operating in the U.S. or anywhere else in the world—the "gold" or economic benefit in this new level of socially sensitive "greening." Companies are and should be moving into sales mode because of new demands and the support or

Preface
Strategic Notes on Corporate Greening in American C-Suites

underwriting provided by government as it encourages less carbon use and more energy efficiency. For example, government money is helping to make hybrid transportation, energy-efficient buildings, alternative fuel and energy generation more affordable to customers, some of whom are under pressure from their customers or local citizens to move into sustainable products and methods.

The company must communicate both economic accountability and social accountability. Management always must do what it takes to sustain its economic viability. Financial performance and engagement with the financial community are job one because without it, nothing else, including social performance, is achievable. First-order work for companies is a communications strategy that has a strong focus on investor relations, government relations, and all the communications that not only assure compliance with disclosure reporting rules but also assure the company is as rationally and agreeably aligned as possible with the stakeholders who are continually deciding whether or not to affirm company management.

Social accountability is similarly important and now, with global warming as a massive central issue, it cries out for expert corporate communications. Sustainability deals with emotional and social matters that may or may not be focused on investment return. This has been cast for a long time as the soft side of company attention. I'm not sure now how "soft" it is. The hard fact is that because of extremely high public awareness, concern and even potential outrage that's being generated by climate change or global warming, corporate communicators are into, and headed for more, challenges in the company's ongoing dialogue with stakeholders. Customers, people in the supply chain, local communities and, very importantly, investors are going to look at the company in a new light and listen to company messages with a new intensity.

Stakeholders are looking at their companies, whether they work for them or invest in them, and they want reassurance and evidence. AT&T's legendary public relations leader Arthur W. Page said the formula for getting public support must start with truth telling, backed up with proof. Former Secretary of State and U. S. Army General Colin Powell has talked about military strategies to deal with uncommon change, something fairly sudden that gets ahead of the normal course of inevitable or expected changes. The ramp-up of climate change is an uncommon change that dramatically shakes up the traditional and established company-stakeholder deal across the board. Market adjustments, reduced-carbon resource management, expectations are shaken or questioned in both economic and social accountability. Stakeholders will look and listen for information that proves the company is sustainable, that it's with the program, that it has the competence to deal with uncommon change, that it can manage the risks, that it can beat its competitors and—bottom line—that this company cares about each stakeholder group's interests and is doing the right things for the right reasons.

The company must communicate with its stakeholders the benefits of its climate change or sustainability strategies. Strategies are ways to reach goals and the goals are sound financial and social performance. Management analyst Jim Collins has written about companies going from good to great, companies managed or "built" to last. Corporate sustainability is—literally, by definition—about that kind of thinking and management. Looking at current business contexts and focusing on a future that will be adjusted and influenced by this uncommon change in fundamentals starting with less reliance on carbon—call it *carbonomics*—driven increasingly by a stakeholder base that is convinced that global warming is a personal threat and a business responsibility.

When polled, Americans show concern, and even fear, about global warming. They expect government and companies to do something.

My view is that there are three threshold choices for a company's climate-change or sustainability strategy: join the leaders, take a middle wait-and-see position, or fight the system. I'd assume very few companies, certainly no public companies, will find it tenable to deny or resist the reasons to form a climate change strategy that deals with risks and goes for returns. Communicators who are in the C-suite as counselors to top management can help other senior executives form the strategy by bringing in the stakeholder interests and impacts.

This is a chance for a company to build trust; chief communications officers are also chief trust officers. Concerned stakeholders look to the company for answers. General Powell made the point that authentic leaders are made in times of uncommon change, when trust is on the line. I like the ad by Perot Systems that says, "trust is earned in

Preface
Strategic Notes on Corporate Greening in American C-Suites

challenging environments." I'd look for the communications opportunity. Move toward the confusion. Learn all you can about what I refer to as carbonomics—that is to say, the change in market economics because there will be a price on carbon emissions and costs that are factored into production and products—to understand and address global warming concerns that tie back to the company. Use what you know to open the trust dialogue with the company's stakeholders. Help lead others in the C-suite to sustainability goals and strategies.

Leading companies are trailblazing the corporate sustainability path that your company can consider. It starts with a clear business case for sustainable performance and communication, making the case that's relevant to the company. You then need buy-in from top management and business operations. This is also a substantial corporate governance issue. When top management is convinced that the sustainability plan is rational and achievable, it will help to get board endorsement. Reports that I'm getting say that boards are approving the first levels of sustainability such as extraordinary commitments to paying for studies and implementation of carbon or energy efficiency measures. Corporate communications can help secure internal company alignment. Information flow needs to be adjusted so the whole company is aimed at sustainable performance. The goal here is workforce involvement, operational accountability, and could mean a culture change. It would be a mistake to neglect this part of communications strategy. You need to get the right kind of commitment, getting the right people on this bus, to borrow Collins' *Good to Great* analogy as you start down this sustainability path.[5] That starts in the house.

Each stakeholder group—investors, employees, customers, suppliers, government—needs to understand, and in the best case be engaged in, the company's strategies. Investor and government relations interactions are particularly important, because the opinions of stakeholders in these arenas can make or break management's resolve. The leading companies are getting allies. I note the value of partnerships with interest groups, government organizations, customers and retailers, understanding their needs, moving toward sustainability goals that they

---

[5] Collins' studies of company success in the books *Built to Last* and *Good to Great* found favor with executives who made it a priority to form effective teams who then collaborate on successful strategies.

can agree on. The U.S. Climate Action Partnership was the primary broad coalition to get business involved and comparatively successful in the 2007 energy bill. Collaborations in most business sectors are active in expectation of congressional and regulatory attention.

The company "going sustainable" needs to rethink corporate contributions. Management certainly needs to ask some questions. What community or regional needs will call for attention as the result of climate change? How can the company get ahead of the curve and make a difference, rather than wait to be asked but to show a leader responsibility? What in-kind goods or services can be provided? How can employees be involved, as they look for assurance that the company cares and is acting toward sustainability? I'd urge attention to localized philanthropy—very strategic, higher impact attention—even while the company is moving globally in its business. Back to the point that sustainability is the combination of economic accountability and social accountability and that contributions are a trust-building opportunity.

Uncommon circumstances are creating common cause. The U.S. Climate Action Partnership, put together to collaborate with Congress on climate change initiatives, includes Environmental Defense, Natural Resources Defense Council and National Wildlife on the same side of the table as allies in addressing this sociopolitical issue. The Prince of Wales' climate change group put something like 150 global corporations in at least a loose alliance with many global NGOs to urge legally binding United Nation agreements to cut greenhouse gas emissions and scale up investment in low-carbon technologies. Environmental Defense is working with Wal-Mart in the very potent influence in green supply chains. World Wildlife Fund joins Intel, Google, IBM, HP, Dell, EDS, Lenovo and Microsoft in the Climate Savers Computing Initiative to set targets for energy efficiency in electronics, and promote global adoption of energy-efficient management tools.

Climate change is providing a share-the-pie experience for environmental groups who understand the economic realities of business and cooperate to put both their green goals and company financials into alignment. TXU Energy, which provides electricity and related services to 2.5 million competitive electricity customers in Texas, becomes the poster child for coal-power utility deals worked out by Wall Street and green activist leaders. Environmental/energy issues were accepted as having financial materiality, and the environmental community was part of the answer for the financial interests. Compatibility will be tested, beginning with roll-out of new energy laws and climate-change legislative jousting. But at the corporate-NGO interfaces, it's not the polarized us-versus-them condition of the old days of going green. Companies need to choose their partners in sustainability plans—and be aware that the NGOs are looking your way and are making *their* choices.

Companies need to examine the value of community or local partners. Cities are becoming very aggressive on climate change. Officials and politicians have to show local constituencies they are with the program. Companies located in, or even selling products in, any large American city already know the pressure will be on them at the local level to show their green credentials, prove their carbon constraint and otherwise help the city "fight global warming." Companies will also be able to look at cities as partners in profitable carbonomics. An unusual opportunity may be in store.

The Toledo, Ohio, story is instructive. Here is a city that's been in an extended economic slide, leveraging its past in the glass business to benefit from the alternative energy demand. Business people see these older industrial cities in a new light, offering clean-tech capability and capacity with the added advantage of community or potential stakeholder support for new plants employing local people and contributing to the economy. This trend, by the way, will mean corporate communications growth, engaging with stakeholders that will be created for some companies in a positive sense by climate change.

The game of winners and losers in sustainability is foolish; the company will win some and lose some. The list of potential winners is long: agriculture, forestry, construction, housing, and finance, consumer goods of all kinds, distribution, and transportation—have I left anything out? The point is that every company will look for its own way to win, to adjust to uncommon change and get ahead of competitors. If you're in the alternative or renewable energy business—biofuels, wind power, even (I consider especially important) nuclear—you know there are putative, fairly near term opportunities. Ethanol—going beyond a reliance on food-crop raw material—shows the potential of fuel sources never previously produced commercially. This can result in hundreds of new factories and facilities, estimates of around a billion pounds of plant material to be raised and hauled around the country, involving trains, trucks and pipelines. Estimates of investments start in the tens of billions of dollars. Sustainable corporations will rise or fall on the bridge of carbonomics. Companies like GE with turbines, power companies moving into nuclear, oil companies into biofuels, automotives into hybrids and flex fuel are placing big bets on climate-change opportunities. Some are already seeing the ROIs that help seal the trust deal with stakeholders, especially company investors.

Coal, gas and oil—front-line targets as government puts lids on carbon dependency—are not instant losers in the alternative energy lottery. If the U.S. is to meet its energy and economic needs for many years to come, it will require these as well as other valuable, available resources. Companies in these industries have moved into the policy leadership to come up with workable answers and regulatory certainty. They are developing the technologies of the future—alternative sources, carbon reduction, CO_2 capture—and working with their stakeholders to meet their energy use needs. All eyes will be on coal burning, which every developed and developing nation relies on as an economic energy source—and which is a huge American resource asset. Technologies to cut coal-burning emissions and to capture them will be developed with government incentives, but other uncertainties cloud the outlook. The FutureGen consortium of big coal producers and users was a highly promising initiative toward coal-fired plants that capture and store its carbon emissions. Like the TXU situation, FutureGen was a test case in the way that carbonomics and hard-to-predict, or manage, external factors affect corporate sustainability. Unlike TXU, the deal (under way at an Illinois site) stalled. Companies and NGOs are learning together how to move through government bureaucracy and other frictions that slow down good sociopolitical decisions—and how critical to forward planning is the level of certainty about regulations, locations and costs.

Transportation will adapt to modest carbon-impact, but this requires transition time. Lighter weight, aerodynamic cars and trucks— some already using hybrid electric technology, a great many running on some percentage of biofuels—are part of the transition now. Government will incentivize this, consumers will accept and then prefer the advanced, alternative private transportation. Heavy highway, school buses and

Preface
Strategic Notes on Corporate Greening in American C-Suites

delivery trucks are becoming carbon light through the use of lower-carbon fuel and batteries. When Amory Lovins, a leading environmental thinker for many years was asked in 2007 by a green blogger if he thought private transportation will remain dominant for the foreseeable future or will there eventually be a shift to public transportation—[such as] high-speed rail. Lovins emphasized transition—doing what can be done over time. He told his interviewer: "We can do a lot better in that regard, with policy and technical innovation, and there are many countries that already do. But with the settlement patterns we have in the United States, it's difficult to make a large shift in a short time in that regard. It's much easier to make the cars, trucks and planes three times more efficient, and that has respective paybacks of two years, one year, and four or five years with present technology."[6]

Global business implications of carbon-era sustainability are huge, a dimension considerably above what we've been through with pollution and greening. The sociopolitical progression of the climate change issue is pitting nation versus nation, developing versus industrialized countries, competing for the right to emit carbon dioxide, which is an indicator of economic growth and viability. This will drive a world of business decisions. Where do you operate? Where is your competition? What does this do to your supply chain? We're already beginning to see impact in China where the enormous supply of goods for export to our country and others will be affected by carbon and environmental requirements. China emits more CO_2 than any other country. Control measures will shake their low-cost structures. Companies worldwide will feel the effects as China and other emerging economies are forced to close, limit or change production to meet greener, more careful consumer and world-trade requirements—and as higher costs to do this are translated into higher prices for exports.

On the road ahead, every company will be affected in some way through rules on carbon controls, carbon pricing, possibly carbon storage. We can already see the carbon war playing out along the typical U. S. sociopolitical scenario—science debates, activism, media interest, politicians or office holders getting involved, one or more crystallizing

---
[6] From the Amory Lovins interview on Grist.org, July 8, 2007

events (such as Katrina was, and such as Three Mile Island was in the nuclear power industry, and Love Canal and Bhopal in the chemical industry), leading to government action. What can you expect? The war on carbon will be institutionalized, embedded in government just as a great many other green and energy issues that have gone from origins, sometimes from the scientific community, sometime from protests and street advocacy, into government bureaucracy. Congress will construct rules of carbon containment and hand out incentives for achievements such as energy efficiencies. Companies will expand their government relations to work for certainties and reasonable terms of compliance with targets and timetables in new regulatory areas. Management, whether a major manufacturer or retailer, whether leading the pack or holding back, OEM or B2B supplier or both, will face operational, technology, product, financing and service questions unlike those of any previous time. Internal physical adjustments in energy use and CO_2 emissions, internal *culture* adjustments (the caring and accountable company culture), will go on while the company responds to market and competitive conditions that are constantly being influenced by customer needs, competitors, citizen activists and concerned investors.

CEOs and CCOs are better prepared to tackle this than with previous iterations of greening. Going green was hard to do, but it's happened. Control of problems and market advantages tied to green and energy issues are institutionalized in most companies. Companies came to terms with gross pollution, implementing control into management structures; call it Greening 1.0. Now in Greening 2.0, environmental managers will get their arms around carbon constraint, and corporate boards will approve sustainable strategic moves and investments. The chief communications officer will help the CEO, CFO and other C-suite executives rethink green strategies by providing intelligence on stakeholder interests, connecting the dots and counseling on positioning the company as a winner in balancing the carbon war's economic, social and political factors. On the corporate sustainability track, the goal for companies is what it's always been: to extend success and gain competitive advantage under changing conditions.

CCO guideposts: *walk green, open kimono, eyes ahead.* Three tips for carbon-era communicators in the C-suite:

Preface
Strategic Notes on Corporate Greening in American C-Suites

One: don't let the "green talk" get ahead of the "green walk." The best PR advice I ever got, from a veteran in the chemicals business, is it's better to be discovered having done something good, so beware the self-anointed green halo.

Two: transparency. Prove it with action, as Arthur Page counseled AT&T executives half a century ago. Tell your stakeholders to track you as a way toward trusting you, as the business leaders at Rio found workable. Report progress (or lack of progress) in achieving metrics: such as carbon footprint reductions. Open your electronic front doors, with everyone invited in, friends and critics. Validate your claims. Get third-party evaluation of both carbon-reducing processes and products.

Three: Don't let the past get in your eyes. Charlie Brown, sage of *Peanuts* cartoons, is told by Lucy that she didn't catch the fly ball in deep center field because she was thinking so hard about all the balls she had dropped in the past. Put aside the rough slogging of Corporate Greening 1.0, where every day there was another ball of hazard—air, water, waste, wildlife—coming at business. We dropped some, we caught some, we were playing catch up. Now look to the new game, taking what we've learned and leveraging it to what can be. Corporate Greening 2.0—with a unifying theme of energy efficiency and carbon constraint—offers scores as well as scars, "green" and the possibility of "gold". Think of aligning mission inside the company with high public awareness of the common challenge. Think of new players on your team, or you on theirs—collaborations, combinations not easily done in the past. Anticipate your future stakeholders. Target the next generation of customers and other stakeholders, recognizing that sustainability is aimed at long-term stakeholder benefits. So think tomorrow. Think Millennium. Know that the kids—tomorrow's customers and employee—are already there. They are ready to believe you can, will, must save the planet and produce abundance. Seems like the right time to get involved in schools and public education about sustainability and the company's commitment to it.

There's little or nothing I'm suggesting here that management and corporate communications chiefs can't do or haven't already done. It's all been proven to work reasonably well, helping management to do its number one job, which is to create stakeholders in the firm's success. It's now a matter of putting this into the new contexts. Understand who your stakeholders are and who they could be, and get engaged in terms that *they* will understand and therefore support. And I could add an obvious footnote, which is maintenance or continuous improvement. This is not a flash in the pan sociopolitical issue. It's building a business that will last, with social and political acceptance. The company will need to create and use systems for internal and external feedback in the sustainability program, to keep it fresh, on track, moving forward in a way that is always relevant to the business and all the stakeholders. Communications people know this is a cardinal practice, and it will be appreciated by the business units.

Corporate Greening 2.0 offers business management teams and especially corporate chief communications officers perspectives and insights that they can use in their work as advisors to top management and engagement with company stakeholders on the current sociopolitical issue of global warming and carbon constraint. I appreciate all the people who have let me in on their challenges and their wisdom. That's what I drew on for the perspectives presented here, and I welcome any comments, corrections or chance for dialogue on the contents of *Corporate Greening 2.0*.

Tips on Climate Change From a Veteran Communicator

Marilyn Laurie has headed both the Public Relations Seminar and the Arthur W. Page Society (the corporate communications group that also elected Laurie to its Hall of Fame, the first woman to achieve the honor). In our conversations on climate change (see Chapter Thirteen), she provided these insights for corporate executives and communicators.

- **Recognize the immediacy.** She sees the move to act on climate change at the corporate level as a speeding cascade of discovery. "Company people have gone very quickly through the ordinary stages of analyzing an issue and arriving at action. The thought process moves from 'this is an issue for the future…affecting the company and my grandchildren at some point' to 'this has immediate potential, affecting my company and my family now.' I follow big trends and I don't recall another developing with such astonishing rapidity."
- **Revisit proven policy.** The former AT&T officer was impressed by the success of the 1987 public and private company agreement on protocols to constrain chemical emissions that threatened stratospheric ozone. "We learned a lot in that exercise. I consider it a very instructive antecedent to corporate engagement on climate change."
- **Stimulate innovation.** Technology developed to address climate change—energy-saving products and production methods, for example—can create customer and revenue opportunities. "The economic value of innovation and the capacity to use it—this is like getting on the right side of the sun coming up."
- **Beware of overstatement.** Communications people need to educate themselves and understand the pitfalls. "Going overboard on green achievements and commitments can be treacherous. Some commitments can lead to unintended consequences—with the obvious example of ethanol."

"The Earth is getting warmer and human activities are a large part of the cause. We need to act in order to prevent a serious problem. The world's best scientists agree we need to reduce greenhouse gas emissions by 60–80 percent by 2050 in order to limit the effects of global warming and this legislation will put us on track to do just that. This is a massive undertaking, and it will not be easy to achieve, but we simply must accomplish this goal; our future and our children's futures depend on it."

—*John D. Dingell, chairman, House Energy and Commerce Committee.* [7]

"As a leader, what you want to do is anticipate trends in the business environment and then position a company optimally within that context. Rather than sitting there and denying that global warming is a problem, my reaction was to accept it and to go with the flow to understand the trend, and then say, how can I turn a challenge into an opportunity."

—*Peter A. Darbee, chairman, CEO and president, PG&E* [8]

---

[7] In 2007, Dingell, Democrat of Michigan and staunch advocate for America's auto industry, supported the first stiffening of federal auto fuel efficiency in 30 years as Congress addressed climate change. The quote is from his statement proposing consideration of a carbon tax in September 2007. More on this in Chapter Nine and in the Appendix.

[8] In 2006, Darbee, whose company operates Pacific Gas and Electric, broke rank with peers to support California's Global Warming Solutions Act, requiring utilities and other companies to make their operations more energy efficient. This quote appeared in an article, "A Law to Cut Emissions? Deal With It," by Jane L. Levere, *New York Times*, October 21, 2006.

Chapter One

Dealing With the Comeback Kid: Five Factors in Play as Executives Zero in on Climate Change

Green is the comeback kid of American corporate sociopolitical issues. C-suite executives who thought for a time that dealing with pollution was pretty much under control are now looking at the deal differently. The many years of deliberate and progressively effective environmental management at the operations level—call it the era of *Corporate Greening 1.0*—now provide the backdrop for a bigger and broader challenge. Top management must now and in the decades ahead come to grips with the sociopolitical conditions arising from emphasis on climate change and carbon emissions. This realization has pushed environment up the ladder of corporate agendas. More than half of the 2,687 chief executive respondents in a McKinsey survey in 2007 zeroed in on the environment, including climate change, as one of the top three issues grabbing public and political attention during the next five years. The implication was that top-level executives worldwide knew that their games were subject to change. They were looking at their business options in a widely perceived necessity to stop global warming.

We are in a period in the world of business that can be called *Corporate Greening 2.0*. It involves economic, social and political factors.

As CEOs and chief communications officers—CCOs—move toward answers within the specific questions affecting their business, it is useful to review the factors that have made "green" the comeback kid and, more importantly, will shape conditions for business decisions in the decade ahead.

Factor 1:
Green has enlisted in the war on carbon.

The old-guard green issue, preoccupied with air and water pollution, waste, recycling, public health and wildlife protection, barged back into the C-suite with new vigor by holding hands with the issue of climate change. The political and social consensus on global warming has given rise to the war on carbon, resulting in two major areas of impact on business:

First: operational. Beyond the ongoing accountability for established environmental matters, companies must get their heads around new linkages of the environment with carbon-connected energy. Power, fuel, products, carbon footprints—a fresh supply of questions is thereby pushed onto the business strategy table.

Second: economic and financial. Among the multitude of new green-plus-black topics that apply to virtually every company is risk exposure re-evaluation, moves to bank carbon credits—cushions against future shocks of emission limits.

Company business communications will need to reinforce the company's social or green accountability in its dialogue with critical stakeholders. CCOs must clearly understand that the carbon war has the momentum to keep green issues alive in new contexts, linked in new ways to essential corporate energy needs, and, at the same time, enlivening some positive prospects for new technologies, products, processes and markets. Management will need to stay close to what's happening in Congress, in the states and in other countries that impact company strategies. Green activism in the form of politics and public policy are in for the long term.

Factor 2:
Green political activism has been strongly revived.

While corporate executives, including communicators, plan winning moves in the new green/energy game, they realize that the game has changed in another way. It has brought onto the political or policy field a refreshed group of players. Like the patient Russian home guard troops in WWII who waited for the weather to change so they could retake Moscow under conditions they favored, green activists in 2007 took back Capitol Hill and quickly flexed their muscles in Washington as

Chapter One Dealing With the Comeback Kid:
Five Factors in Play as Executives Zero in on Climate Change

well as in key states.

When the Supreme Court subsequently ruled that carbon dioxide could go into the basket of regulated pollutants, when California rolled up its sleeves to lead the new wave of black-plus-green controls (a lead that at least 15 states said they will follow), the course was set for environmental policy making of unprecedented intensity and staying power.

By the time of the 2008 presidential and congressional elections, as in no previous election period, there was a collaboration of attention on a central environmental issue. Climate change became an organizing principle in virtually every area of social accountability. Eco-activism became mainstream. Political hammers began nailing down the future for business and its green/climate challenges.

Environmental groups, the news media, politicians in and out of office and federal and state regulatory authorities are now energized and looking to the business community both as part of the problem and as a substantial part of "the answer to global warming."

Factor 3:
Pressure is mounting from investor activists.

In addition to sociopolitical activists, there are new green socioeconomic activists. Main stream investors and customers are demanding corporate insights on the potential for global warming to burn an investment portfolio.

Wall Street, pension fund managers and others invested in corporate well-being have put their spotlights on companies positioned to capitalize on carbon-war conditions; and they are probing for vulnerabilities in others, raising questions such as these:

Is the company a winner or a loser as carbon is constrained? Can the company withstand physical climate change, energy cuts, environmental responsibility, social and cultural demands? Can this company's executives beat the competitions'?

Gadflies from the first green wave who stung companies to move on pollution cleanup are meanwhile reinvigorated by global warming. As an example, Ceres, the corporate governance hawk that was born during the time of acid rain and the Valdez oil spill, now challenges corporate management to disclose climate change risk that investors should factor in. "It's no different from litigation risk or hazardous-waste risks or anything else disclosed in normal financial filings," said a Ceres official, indicating the level of inquiry now on C-suite desks. Investors filed a record 54 global warming shareholder resolutions with U.S. companies during the 2008 proxy season—nearly double the number filed two years previous—despite the fact that the number was held down somewhat by companies agreeing to eco-friendly moves in advance of the proxy season. With climate change now rooted, groups such as the Carbon Disclosure Project, a not-for-profit group with more than 300 institutional investors including Goldman Sachs and Merrill Lynch (among the several major firms that have set up green business practices focused on climate change), have pushed companies and their business partners to reveal carbon footprints in a consistent quantitative way, to enable investors and the public to evaluate the companies' sustainability. This is expanding to measure carbon risks and liabilities in company supply chains. The data-gathering and evaluations process is awakening companies of all sizes to the inevitability of total transparency related to climate change.[9]

Business executives, CEOs and CCOs, must now answer a question posed by a growing base of old and new stakeholders: What is the exact nature of our new green strategy, now that climate change is in play?[10] What are the dynamics of Corporate Greening 2.0, with its focus on energy and climate threats, how can we leverage what's been learned and achieved in managing the large, common environmental challenges of Corporate Greening 1.0?

Factor 4:
Green-collar executives can and will make a difference.

Surveys and the media have brought ample evidence of global C-suite awareness of the challenges. A burst of activism in Washington in 2007 and 2008 established a sociopolitical front for the new impact of environmental accountability. Across the U.S.—in Sacramento and many other

---

[9]The Institute of Management and Administration (IOMA) provides the results of its extensive survey of supply chain reaction to climate change demands in its "Guide to the Carbon-Efficient Supply Chain." More information at www.ioma.com/supply_chain/.

[10]Corporate investor pressure and its implications are discussed in greater detail in Chapter Seven.

Chapter One Dealing With the Comeback Kid:
Five Factors in Play as Executives Zero in on Climate Change

state capitals—executives, and often chief executives—were engaged in the climate-altered process of public policy formation. The unusual green collaboration known as the U.S. Climate Action Partnership put CEOs and green activists into the earliest rounds of congressional negotiations on energy and climate change legislation. Older line executive conclaves such as the Business Roundtable, the U.S. Chamber and the National Association of Manufacturers have wrestled with ways to adapt their business-based focus on public policy to deal with the changing green dynamics. While, as McKinsey's research showed, chief executives have increasingly incorporated environmental, social, and governance issues into core strategies, many of them are increasingly involved in climate-change public policy.

What motivates this involvement? McKinsey's pollsters found a sizable group of executives personally worried about global warming and presumably motivated to rethink green policies out of fear. My sense is that for most C-suite executives, the greatest fear is not about change, climate or otherwise. It's about uncertainty. That is the motivator that puts corporate leaders into policy formation, to un-

TIPPING POINT FOR GREEN RESOLUTIONS

While it has not been uncommon for investor activist groups to withdraw proposed resolutions when companies comply with their terms, 2004 saw an increasing number of these events as climate change moved up as a sociopolitical issue. Under pressure from shareowner activists, energy companies Cinergy, American Electric Power, TXU, and Southern Company all agreed to prepare reports on the risks posed by climate change and company plans to mitigate such risks, and Reliant agreed to increase climate risk disclosure.

While these green resolutions were not voted on, the Bank of Montreal that year became the first company in Canadian corporate history to recommend voting for a social or environmental resolution, after activists asked the company to disclose how it evaluates and manages environmental risks to its business. The same year, Tyco backed a similar resolution asking for a report on its corporate-wide toxic emissions. The company-endorsed shareowner resolutions all received near-unanimous support (91 and 92 percent respectively).

Source: SocialFunds.com

derstand social and political forces, to grasp the prospects of business accountability, and to negotiate the terms of change. In short, the best outcome will be one that is effective and achievable.

A significant example of executive acceptance of changed social and political conditions has occurred in the American auto industry. Long a target of green activism against fossil-fuel use, company leaders became actively engaged in the federal policy process on climate change with the benefit of strategic political guidance from the veteran lawmaker John Dingell, Democrat of Michigan, who came back as chairman of the jurisdictional committee as the result of the 2006 congressional elections. Understanding the realities of politics and markets, sensing the prospect of customer as well as political support for advancing the required technology, Ford's top executives went green-proactive. They agreed to tough future rules on fuel efficiency, assuring some level of certainty as to what is expected of them, and releasing them now to work with suppliers, customers, dealers, unions—in short, all the company's stakeholders—to move toward compliance. There was a move to mobilize around the idea that policy makers who want autos to do their part to answer the global-warming challenges will assist business by favoring technology advancements, tax and trading rules that surround the prospects for clean, green vehicles.

USCAP, led by some of the same companies that wrestled with government policy from positions of weakness during the early waves of Greening 1.0, started its drive to shape regulatory expectations by okaying mechanisms like carbon caps. Similarly, before national leaders locked in on green policies at the United Nations climate conference in December, more than 150 global companies signed on to a call initiated by England's Prince Charles for mandatory greenhouse gas controls. Dozens of industry-specific organizations such as the Edison Electric Institute[11] were quick to engage in the congressional policy process with positions favoring government action shaped with them at the table. These and similar moves led by Western, primarily American corporate executives have led to some sense of common expectations at a time of uncommon conditions.

To summarize, the green comeback kid is considerably more muscular with energy and climate change now added to its traditional environment orientation. As McKinsey found, this newly raging

---

[11] See Appendix for more on EEI's advocacy position.

sociopolitical issue has awakened executives not to retreat but to engage. Especially in North America, where democracy provides a bowl match for this type of policy game, corporate leaders are in a pragmatic green-collar stance, agreeing to certain mandates to mitigate the dangers of climate change while working with lawmakers toward achievable, market-based compliance, reliable long-term carbon price signals, regulatory certainty and economic safety valves.

FACTOR 5:
CORPORATE COMMUNICATIONS MOVE TOWARD SUSTAINABILITY.

Respondents in the McKinsey survey identified environmental concerns as the most important trend influencing public expectations of business. With stakeholder expectations in the wheelhouse of corporate communications, this puts CCOs in an important carbon-war role.

Company communication strategies, methods and tools must engage with specific, critical stakeholders to determine exactly what are their concerns and expectations. This starts with research into make-up and attitudes of specific stakeholder groups, a process necessary because of the heightened awareness of global warming as an issue of major concern and of the changes occurring in markets and financial circles. In major large companies, communications are now being approached with a corporate sustainability mindset.

Corporate sustainability means essentially a workable, synergistic approach in addressing social expectations while succeeding at the main goal of any enterprise and that is satisfactory economic or financial performance. The sustainability mindset says that social accountability—in this case, social impact resulting from climate change putting energy into green contexts—has taken on full acknowledgement of economic accountability. It acknowledges that companies are judged first and foremost for their financial performance. In his incisive, pragmatic book on "the halo effect," Phil Rosenzweig underscores the fact that good scores ("halos") chalked up by a company on social, cultural and other accounts are almost always just an extension of the good scores on the financial front—profitable products and operations, return on investment and the

---

[12] *The Halo Effect*, by Philip M. Rosenzweig, Simon and Schuster, 2007.

like.[12] With sustainability, as I interpret it based on conversations inside companies as well as studying what experts and major players are saying, social accountability meets the hard road of economic accountability. Each enables the other, but social good without financial soundness is a frail foundation for enterprise endurance. This is not to suggest that what we've been calling "corporate social responsibility" is abandoned.[13] In fact, the socioeconomic melding that's forced by climate change should mean an elevation of CSR as a substantial C-suite consideration. CCOs and executive level peers who have found any resistance to bringing up "CSR" as a serious agenda topic may find it getting more respect when it's part of the corporate sustainability agenda. In this book, I will take this one step farther. I will bring politics into the mix for the simple, and simply profound, reason that companies must be set on the sustainability course as the result of government. If politicians and government officials did not recognize the necessity for all-out war on carbon, did not declare that carbon dioxide is a pollutant to be regulated, did not embark aggressively to change current market economic factors into what I call carbonomics—then C-suite discussions about social good and environmental protection and energy factors might still be separable. My contention is that the silo days are done. Greening 1.0—clean it up, manage it, pay the price—is supplanted by a new tripartite accountability that is social, economic and political: Greening 2.0, driven by climate change and energy interactions.

This is the new sustainable communications platform that puts chief communications officers at the C-suite strategy table as well as in the various field positions of identified stakeholders, ready to address the impact of climate change and to contribute their special insights, gained from their study of stakeholder instincts and interactions, to the company efforts in the next levels of corporate greening. I won't at this point cover the evidence of strategic moves under way to advance corporate sustainability. That is offered elsewhere in this book.[14] Let me just observe now that a great many corporate Web sites (many now rebranded as "sustainability"), investor relations, executive speeches, interviews at shows and publications of various kinds are coming into alignment to

---

[13] CSR is firmly established as a business function and service practice, backed by scholarly research and academic studies. This is comprehensively demonstrated in Andrew Crane's *The Oxford Handbook of Corporate Social Responsibility* (Oxford Handbooks in Business & Management), published in April 2008.

[14] See Chapter Ten: Communicate Your Carbon War Strategies.

Chapter One Dealing With the Comeback Kid:
Five Factors in Play as Executives Zero in on Climate Change

deliver the company sustainability pledge: *Our business, adjusted for climate change, is sustainable. Here is our record, here is our evidence, and this is how we will move forward with our stakeholders' interests in mind.*

With internal and external evaluators looking at the company through lenses changed by their perceptions of global warming, communication recalibrations are in order. The chief communications officer's role is as always about two things: managing information flow and managing expectations. Information must get to the right place at the right time. It must be transparent and bidirectional. Expectations inside the company and with all stakeholders must be as close to realistic as possible. Stakeholder approval has to be earned, over and over again, always starting with what each stakeholder is expected to approve. Senior executives generally know by now that "PR" can't fix bad judgment and loose commitments. In the near-term phases of the war on carbon, the best achievable outcome for corporate communicators—who I consider to be stewards of stakeholder trust in the company—is to help others in the C-suite rethink green so when decisions are made the message is true, consistent and understood by the stakeholders. Greening 2.0—with energy joining and leading environmental factors as the result of climate change—is the new sociopolitical front. In the house and along the company's external interfaces, strategic business communication will help shape the odds for success.

Green Technology as the New Y2K

Back in the late 1990s, when it was not clear precisely what it would require to reconstruct enterprise computers in advance of the date change at the turn of the millennium, the financial-services and corporate worlds bit the bullet and redesigned their technologies.

The global Y2K effort may or may not have prevented global computer collapse; we'll never know. But its effect was immense either way. It spurred innovation, it promoted cleaner housekeeping, and it may well have lessened the impact of disasters such as the September 11, 2001, attacks—if only by making the financial services community more aware of its capacity for safeguarding against breakdown and recovering from it.

With the threat of climate change, the governments and corporations of the world face a new Y2K-style challenge. Whatever the actual threat of human-created greenhouse gases may be, the perceived threat—and the urgency among groups like the Intergovernmental Panel on Climate Change—is going to drive both governmental and corporate activity for the forthcoming years. This will lead to a wave of housekeeping and innovation that, like Y2K, will have unexpected effects.

This does not necessarily mean that conventional wisdom (for example, about the immediate returns from energy-efficient technologies) is right. But capital investment priorities will change, many companies will develop innovation or alliance skills that they otherwise would not, and waiting for government mandates will probably not be the most effective strategy.

Source: Strategy+Business, *published by management consultancy Booz Allen Hamilton, included this among eight management trends cited in an article, Signals for the Future, by Art Kleiner, February 22, 2008. www.strategy-business.com*

Profit in Energy Efficiency[15]

Cuts in greenhouse gas emissions to make the world safe from global warming can be achieved at a net profit to the global economy, according to a 2008 study by McKinsey consultancy. The study concludes that investment in energy efficiency of about $170 billion a year worldwide would yield a profit of about 17%, or $29 billion.

Mindy Lubber, president of Ceres, a coalition of institutional investors that commissioned the report, said: "This is the message financial leaders need to hear: there is huge opportunity [in energy efficiency] and those moving money around are going to make the difference in this. Efficiency is the fastest, cheapest way to reduce greenhouse gases and could bring large profits to the global economy." Lubber said institutional investors are seeking more energy-efficient property portfolios.

---

[15]"Study finds profit in cutting emissions" by Fiona Harvey, *Financial Times* 2/14/08; copy of the report available online from McKinsey.

"We have become convinced that modern environmentalism, with all of its unexamined assumptions, outdated concepts and exhausted strategies, must die so that something new can live."
—***Michael Shellenberger and Ted Nordhaus***[16]

"Some of the most powerful corporate leaders in America have been meeting regularly with leading environmental groups in a conference room in downtown Washington for over two years to work on proposals for a national policy to limit carbon emissions. The discussions have often been tense. Pinned on a wall, a large handmade poster with Rolling Stones lyrics reminds everyone, 'You can't always get what you want'."
—**New York Times** *article by* **Jad Molawad**[17]

---

[16]From their essay, "The Death of Environmentalism: Global Warming Politics in a Post-Environmental World," published on the Web, October 2004.

[17]"Industries Allied to Cap Carbon Differ on the Details" appeared in the *Times* on June 2, 2008. It describes workings within the U. S. Climate Action Partnership among business executives (Duke Energy, Ford, ConocoPhillips, FPL Group mentioned specifically) as they and their green group partners (Natural Resources Defense Council identified in the article) came to grips in 2007-2008 with congressional action on climate change legislative options, especially the Senate's Lieberman/Warner bill that led the debate on cap-and-trade factors.

Chapter Two

How Business Arrived Here:
The Hot and Cold Trail of Sociopolitical Greening

Much like the Earth's warming and cooling periods noted by scientists in the debates over climate change, the strategic value of greening in political and business arenas has gone up and down. We are presently in a warming period.

Arguably, it's the hottest level yet. Largely because of the fusion of climate change and energy matters, environmental issues are ranking with or above the high level of public awareness and heat traceable to the 1962 publication of Rachel Carson's book *Silent Spring*. In the years following, anxiety and outrage about gross pollution, mindless waste and threatened species rose to produce the equivalent of a morality drama. American business was commonly cast as the villain or antagonist, confronted by protagonist environmentalists and politicians who provided the climax of the drama by enacting laws regulating business behavior on behalf of victimized people and nature. Environmental organizations grew rapidly. Backpackers became urban protesters. New groups sprang up, some community based, others with national and even global aspirations.

The result in this period—call it Greening 1.0—was not only the transformation of conservation to green activism, but the start of a collaborative move upward, with American capitalism joining with environmentalism in support of public good, contributing essential money, management competence and innovative technology. Thermometer readings of this beneficial move upward, however, would appear jagged.

WARM SUPPORT FOR, OR DEATH OF, ENVIRONMENTALISM?

When Georgia Governor Jimmy Carter became president in 1977 after waging a strong campaign with environmental protection as a central focus, I said in a commentary published by the Public Affairs Council that this marked the end of the environmental movement. For this opinion, I got some bashing from environmental writers, including fellow members of the Society of Environmental Journalists, in which I was active from its beginning. This was certainly understandable if for no more than the fact that the Public Affairs Council carried the title "Death of Environmentalism" over my musings.

The thesis I presented was that the success of his green supporters in taking Carter to Washington would effectively take the activists off the streets, where they had been advocating government (as well as business) action, and put them into government. They could propose and probably legally require at least some of the actions they had touted as outsiders. When a movement reaches the pinnacle of power, went my reasoning, it ceases to move. It stops agitating. It starts to give orders. Rallies become rhetoric—and rules. In that sense, I was not wrong to note the termination of the organized green movement, and the victory of institutionalizing within government—an outcome which over the course of four decades has produced the heaviest burden of social laws ever imposed on business operating in any country.

LIBERAL ENVIRONMENTALISM FELT DEATH IN 2004

Little did I know then of course that 17 years later, a young leader in the green movement would give a speech at the famous Commonwealth Club in San Francisco that echoed my topic, but took an entirely different tack. Adam Werbach, who had founded the Sierra Student Coalition and in 1996, at the age of 23, had been elected as the youngest-ever national president of the Sierra Club.[18] In the autumn of 2004, he stepped up to the rostrum to announce that he was done calling himself an environmentalist and told his startled audience of fellow greens that "I am here to perform an autopsy." He broke into Latin—*"Hic locus est ubi mors gaudet succerrere vitae"*—and helpfully translated: "This is the place where death rejoices to teach those who live. With fond memories, a heavy heart and a desire for progress, I say to you tonight that … environmentalism is dead."

Werbach went on to explain that he was terminally depressed over the success of conservative forces in American sociopolitical circles. "Our death," he said, "is a symptom of the exhaustion of the liberal

---

[18]In 2003, Werbach was appointed by San Francisco city supervisor Chris Daly to the Public Utilities Commission while then Mayor Willie Brown was out of town. In 2006, he began to work with Wal-Mart to help lead their efforts in sustainability; in the same year, he was elected to the six-member international board of Greenpeace. (Source: *Wikipedia*.)

Chapter Two How Business Arrived Here:
The Hot and Cold Trail of Sociopolitical Greening

project. Having achieved its goals of basic economic rights, liberalism and its special interests now fail to speak to the modern need for fulfillment of the American people."

For eight years, Werbach said, he had been trying to tell leaders of the environmental movement about the severe impact of these somber facts, and now was giving up; he had failed to communicate effectively. Werbach's mock funeral for green activism touched off a heated discussion inside the community. Much of this centered on the environmentalists' view of national politics since the 1994 elections when a less-liberal, more-business-friendly Congress and soon-to-be Speaker Newt Gingrich's[19] "contract with America" was perceived to sideline their agenda and influence.

Werbach's dismal viewpoint was reinforced by two other well-known green thinkers, Michael Shellenberger and Ted Nordhaus,[20] who wrote a memo widely read when it was placed on the Web. Further detailing Werbach's case, they spelled out a syllogism of three arguments:

(1) that environmentalists were making inadequate progress on global warming;
(2) that they had not mobilized public concerns and values to create political pressure to make changes in the way society uses carbon; and
(3) this inadequacy is because of progressive social movement failings, contrasted with more strategically integrated efforts of conservatives.

Others inside greening pushed back. In December 2004, Sierra Club Executive Director Carl Pope distributed a long letter to environmental grant-makers responding to these views, acknowledging that the state of the movement was challenged, but arguing that the antidote was not surrender but strategies to fight back. Pope and other green group

---

[19]Rep. Gingrich was Speaker from 1995 to 1999. *Time* Magazine selected him as the Person of the Year in 1995 for his role in leading the Republican Party revolution that ended 40 years of Democratic Party majorities in the House of Representatives. Following the 1998 elections, when Republicans lost five seats in the House, Gingrich announced his resignation as Speaker.

[20]In October, 2004, Shellenberger and Nordhaus, well known in the green community as strategists and consultants, authored a controversial essay, "The Death of Environmentalism: Global Warming Politics in a Post-Environmental World," which through the Web quickly became a hot topic. Picking up on Werbach's theme and zeroing in on climate change, the authors concluded that the "institution" of environmentalism was now incapable of dealing with climate change and should "die" so that a new politics can be born. (Sources include *Wikipedia*.)

executives still felt the sting of the Gingrich revolution, now a decade old and considerably changed. Pope's friends in the news media remembered his words to them after the Republicans' win in 2004; in effect, *Just wait. Our time will come again.* He indicated that the greens and the Democrats would come back, much like the Russian troops in World War II who held off the German troops outside Moscow until winter arrived to favor the Reds and drive off the intruders. Pope and his colleagues rose above the towel-tossing started by Werbach. They sought continued support for organized green efforts and looked to new potency from global warming as an engine for liberal progress and American environmentalism.

Transformational Change

But back now to 1997. President Carter is in the White House; the greens, in the federal bureaucracy. Under Carter, the U. S. Environmental Protection Agency, which had been set up by President Richard Nixon in 1970 following congressional authorization, began to broaden pollution control from a focus primarily on natural resources to a focus on health. This would provide environmental advocates, whether in government or in the field, with new lines of attack, complicating the response of company engineers and environmental officers to government's green requirements. Rachel Carson had pointed to pollutant effects on bird eggs; new green causes would be grounded in research and claims on suspected human health effects. Stretched from the realms of wildlife, endangered species and habitats to virtually all human encounters with air, land, water and waste, the blanket of federal environmental regulations grew to cover virtually every private sector endeavor.

With Carter and his green forces in place, the echoes of protest and public demonstrations for green change became background for the newly acquired front lines of green action—in the halls of government as well as in the offices of businesses large and small where the hammers of regulations would fall.

The environmental movement had come to a place of power. While the Carter presidency represented a transformational change in American socio-politics, while green activism had gone from street to bureaucracy and while the blanket of regulatory coverage had been enlarged from clean-up to health, it was certainly not the death of environmentalism. Far from it. In fact, we were at the door of another transformation that would set the stage for the present world-wide war on carbon.

Chapter Two How Business Arrived Here:
The Hot and Cold Trail of Sociopolitical Greening

Temperature Rising in Rio

Since 1992, when the world awakened to the sociopolitical theory of "sustainable development," greening has steadily gone global. Man's impact on the environment, economic development as a matter of deep concern, and the prospect of controls put under new international treaties were the focus points of the first-ever United Nations conference on sustainable development. Held in Rio de Janeiro, the so-called "Earth Summit" was the big tent for big green talk: the leaders of nations, large and small, come together to discuss international treaties on environmental development. World interest was huge. More than 100,000 people were drawn to the event. The conference would serve as a substantial kick-off for serious environmental responsibility in many developed countries, and green rhetoric on a world stage would reach fever pitch.

It was the largest gathering of green activists ever. An estimated 2,400 representatives of non-governmental organizations were among the registrants; another 17,000 people attended the parallel NGO Forum with its open-air booths, demonstrations, speeches, games and programs celebrating the opportunity to express environmental views and values.

And most of all, it was a magnificent media event. Among the participants in the main event—the formal proceedings of the 172 nations, including 108 heads of state—were 8,000 conference participants who signed in as "journalists." The generation of massive news coverage boosted (and for many people, introduced) the concept of "sustainability" as an organizing principle for environmentalism.

Leader after leader rose to testify for planetary actions. Under-developed nations' leaders described their plight and the developed nations, their fight—or their intentions to fight—against degradations in nature. Media observers saw it as a contest between have-nots and haves, with the 77 have-nots (or have less) nations scoring the most points. Ironically, the nation that had invested the most in environmental cleanup, had erected the most widespread government control system, and had the greatest potential to move toward realistic world greening was the nation most abused. The United States was brought onstage in a negative, largely defensive position.

It had been made clear, through media coverage and pre-conference sessions as the U.N. event neared, that the U.S. was serious about its green initiatives, intentions and willingness to work with other countries—but would not be pushed into any agreements that would inequitably punish the nation's economy. President George H.W. Bush came to Rio, bravely I thought, and took the blows of criticism for the U.S., contributing by his presence the leading nation's commitment to the sustainability cause—but leaving with only slight engagement in the rhetoric inside the U.N. tent.[21]

OFFSTAGE, COOL HANDS OF AMERICAN BUSINESS

While the hundreds of political and government leaders gathered in the great hall to debate treaty proposals, a handful of CEOs in a separate session put forward a new commitment from the business community, a challenge to those doubting corporate commitment. Business leaders from Western nations, led by American corporate executives, took advantage of the Rio conference to raise the green flag of private-sector commitment. Senior company executives and their staff environmental and communications people (a group in which I participated) advanced the newly-formed World Business Council on Sustainable Development. The message provided personally by CEOs from major American companies and European counterparts was about engagement with stakeholders, government and NGOs at new levels of transparency. Essentially, the corporate message was this:

We understand that people are concerned about the environment and that business is often cited as a major part of the problem. We're no longer asking the public to "trust us" when it comes to the environment and what we're doing about it. We're saying we will be much more open than we have been in the past when it comes to information about our environmental management inside our various companies. We will provide hard data to show our progress, or lack of it, in controlling air, water and waste pollution. We will start with a base year and give you annual data, figures, an accounting of what we're doing in terms you can understand and analyze. We invite your inspection, your questions and your help as we go forward together.

The high-level delivery of this message—boiled down to "we're not asking you to simply trust us, we're asking you to track us"—didn't

---

[21] My commentary on this episode and a comparison provided by a 2006 U.N. event in which President George W. Bush participated can be found on page 39, "Bush Avoids Dad's Dilemma."

raise a lot of interest on site in Rio among the but it was significant on several counts. American companies who had been under the green fire (some of them for 30 years, since 1992 was the 30th anniversary of the *Silent Spring* attack on chemical products) got grounded in this sociopolitical issue at Rio. I consider this to have been a turning point in corporate green communications, using the confidence gained from successful environmental management under three decades of intensely pressurized conditions to go from defense to pro-activity.

BIRTH OF SUSTAINABLE BUSINESS COMMUNICATION

From the timeframe and events of Rio, as the age of green accountability unfolded with tracking and transparency as its guidelines, business communicators were positioned more substantially as primary links to company stakeholders. Sustainable development, centerpiece of the Earth summit and motivator for subsequent moves to adopt international environmental treaties, made sense for business as well as for governments. Like a tree with many branches reaching for the future, the concept of "sustainability"—now a staple of corporate green communications—was rooted.

GREEN NEWS FROSTY

If you jump forward a dozen years, the tree has grown and the heat of public opinion about the environment is in something of a cooling period. It is 2004, an important presidential election year and environment is not a winning topic. Green activists fail to attract reporters to their messages of conservation. The emerging issue of global warming, while hotly debated in the scientific community, finds a frosty reception at the grassroots realms of politics.

What were the reasons? Voters had other concerns. Except for limited interest in a few blue states, American voters were relatively cool to green as a sociopolitical issue. The George W. Bush-John Kerry debates found no voter appeal for green charges and pledges. Distractions of war news (9/11 and its aftermath affected everything) helped drive green toward the bottom of public surveys. During the final campaign weeks of 2004, a Gallup poll placed environment lowest among current national concerns, falling to relative obscurity as a subset of energy policy, which itself ranked far behind terrorism, the economy, Iraq and healthcare.

Another factor figured into reduced public perception of environment as a condition for public concern: *the impact of business communication*. A good part of the reason that the greens could not ramp up the political rhetoric against pollution was that business had done enough and communicated well enough to move generalized public perception toward a view that pollution in the U.S. was under better and better control.

Major concerns about acid rain, ozone layer destruction, ambient air, polluted rivers and garbage were mitigated by evidence of improvement, reported by the media and by companies involved as part of the answer in each of the challenges. As promised, American companies had taken accountability for their green impact. They were providing measurement metrics and working as partners with stakeholders including NGOs. Though still under the watchful, often skeptical eye of government and NGO activists, companies were not so intensely under the "fierce green fire" that Phil Shabecoff described in his book on the environmental movement. Business had developed confidence in environmental management.

Companies had launched green initiatives and their communicators were comfortable in dialogues with stakeholders about these. At the investor level, climate change rose on the agendas of social activists, and shareowner resolutions demanding company disclosure on greenhouse gas emissions and pledges to cut carbon.[22] On the 2004 political trail, however, there was very little anti-business green talk. The overwhelming themes were foreign policy and the war on terrorism, centering on the U.S. engagement in Iraq since 2003 invasion of Iraq. National security, the economy and health care were issues more likely with voters than the environment or climate change and business.

"SPRING" IS WARM FOR GREENS

Two years later, the shift in the public story was dramatic. While the environmental community's candidate, former Vice President Al Gore, had not reached the White House in the 2000 elections, his party and his

---

[22] As resolutions on climate change and other social issues rose, however, several companies cut off the contention by agreeing to shareowner calls for action. "The shift from confrontational relationships between shareowners and corporate managements to more collaborative ones has been years in the making, and 2004 saw a sea change in tangible results and innovative solutions for this new, mutually beneficial approach," commented Jay Falk, president of SRI World Group, which publishes the SocialFunds.com web site. More on this in chapter seven.

Chapter Two How Business Arrived Here:
The Hot and Cold Trail of Sociopolitical Greening

environmental supporters were able to celebrate the Democratic success in Congress in November 2006. Carl Pope's vision of the Russian winter, with the home troops regaining the field, had come to pass.

What had caused the shift? Public issues go through phases. One scenario is the science to government scenario. An issue bubbles along in the scientific community. An action group or two pick up on the issue and bring it to the attention of the news media. Politicians are intrigued by the issue and see the possibility of engaging to help constituents and create voters. Then—something happens to crystallize the situation—an event, a Three Mile Island, a Love Canal, a Bhopal. The event ties it all together and government takes it to the finish line. Activists, media, politicians and others—business is not excluded if the conditions are suitable—collaborate and laws are passed and regulations are rolled out.

Sometime in the first half of the first decade of the 21st Century, the sociopolitical issue of climate change traveled along the science to government path and set the nation into the extended war on carbon. The long-running debate in the science community about global warming—its magnitude, causes and projected effects—had been overtaken by events that would assure climate change as the century's leading sociopolitical issue. A hurricane called Katrina, a tragic weather event linked by the media and others to climate change, an indefatigable Al Gore staying on a message of planetary peril caused by climate change, and the return of Democrats to the control of Congress with a pledge to act on climate change: these combined to drive green back toward the top of the charts. The perceived value of green news had risen, as the direct result of putting the science of global warming into social, political and business contexts. That is the condition as this book is written.

FICKLE POLLS AND SOCIOPOLITICAL EFFECT

I want to underscore here that even as issues climb in public awareness, we are not talking automatically about widespread general heat for political action. While surveys of American public opinion over the years generally favor green action, green products, green corporate responsibility, they are not overwhelming mandates for government or business reaction. Furthermore, polls—or people responding to polls—are fickle. A large number of people may respond to a survey saying they are green, are going green, want to go green and will advocate for companies to go green—and then go about their lives pretty much as before, expecting government or somebody else to do the greening they say they favor. Opinions matter because people who tilt toward green can be led to support green action. Followers generate sociopolitical leaders—activists, media, politicians.

We are talking about vocal minorities, aggressive spokespersons and superconductive communication channels—from bloggers to TV casters—pushing the issue into the realm of politicians who move it toward government mandate. That is when the war guns get trained on business.

Polls of American public opinion on green subjects have over the years shown few spikes in general concern and little evidence of personal action. An ABC News poll at Earth Day 2007 revealed that most people consider global warming the world's most serious environmental problem—94 percent of them said they were willing to make personal changes to help solve the problem. But will they? As Joel Makower has pointed out, eight in 10 Americans in 2007 also said they oppose increasing taxes on electricity to encourage energy conservation, and about two-thirds said they oppose more taxes on gasoline and raising prices at the pump. A good-sized poll by George Mason University in 2008 found just a slim majority of Americans who considered global warming a very serious problem.[23] Concern about global warming has only modestly increased in surveys done over the past decade at Stanford University.[24]

Two points seem quite clear. First, this issue—green action and climate change as a motivating focus—has legs and will energize media coverage, elected officials and social activists. There are concerns, there is news, there are constituents, and there is a social cause. The war on carbon is already beginning to settle into sociopolitical accommodation that will have long-term business effects. Carbonomics—rethinking green economics to factor in the costs of carbon constraint and the return from carbon control opportunity—is on management's table. And, second, for the first time in the more than three decades of green issues, business has taken a substantial early role in shaping the outcome.

---

[23]Poll of 11,000 Americans showed 62% considered global warming a serious problem, with this opinion much higher—at 78%—among those who "always" identify themselves as Democrats than the 36% level of those who are "always" Republican. Edward Maibach, senior author of the survey and head of a center on climate change and communication at George Mason University in Virginia, told *USA Today*, "Clearly that's a lot left to do in raising awareness." Source: *USA Today* article, "Actions Don't Match 'Green' Attitudes," January 31, 2008.

[24]Jay Krosnick, Stanford University political scientist, quoted in the *USA Today* article, as saying "(People are) more enthusiastic about government and business doing things than doing stuff themselves."

Chapter Two How Business Arrived Here:
The Hot and Cold Trail of Sociopolitical Greening

BUSH AVOIDS DAD'S '92 DILEMMA

Commentary: June 1, 2007—The surprise package on global warming that George W. Bush (Bush #43) opened before he went to the G8 meeting in Germany to face the issue in an unfriendly forum on the subject reminds me of what happened to his father in a similar situation 15 years ago.

In June of 1992, the forum that George H.W. Bush (Bush #41) faced was the first so-called "Earth summit" held in Rio de Janeiro. The subject then, as now, was what nations' leaders would do, or commit to do, about nagging environmental matters. Then, as now, there were pending matters related to agreements among the national parties, including climate change.

At the United Nations conference back then, there were half a dozen draft agreements—conventions, they were called—that could lead to treaties.

The current President Bush made a considerably smoother strategic move than his father on global warming.

Biodiversity, sustainable development and global warming were on the table, ready for action by the chiefs of some 77 countries. The buildup to the main event had been long and intense—three years of pre-conference sessions of various underlings juggling ideas, debating the meanings of sentences and words, sorting out what all of it means to every country, developed and undeveloped. Translations slowed the process, the ground was pawed by the delegated emissaries, until finally the drafts were done, and it was show time for the people with real power, their bosses.

One sports columnist said at the time it was shaping up into a north-south football game. Only, the teams were mismatched. The north was the G7 group (now G8) of large, developed nations, competent, proud, and accustomed to winning. And the south was the G77—a scrappy group of smaller, less developed, or emerging nations, roaring onto the field that they owned—South America—and ready to rumble.

It was clear as the UN conference neared that the predominant mood of the have-nots was to take what they could from the haves, and that the U.S. was not going to be pushed into any agreements that would inequitably punish the national economy.

So, Bush #41 knew what he would walk into when he got to Rio—at least a cool reception, or maybe a firestorm.

Before he left Washington, the president sent his head of the Environmental Protection Agency, Bill Reilly, to take some of the heat. Reilly tried. He set up pre-meeting interviews to explain how the U.S. was tackling environmental issues since EPA was set up 20 years earlier (far before any of the other nations seriously institutionalized greening), and our intention to partner with all the nations to come up with rational world environmental solutions. The 8,000 conferees registered as journalists had little interest.

In livelier pre-game warm-up rooms, there were Ted Turner and Jane Fonda. Jacques Cousteau was there, as were Bella Abzug, Jerry Brown, the Dalai Lama and former Agent 007, Roger Moore. John Denver sang, and Bianca Jagger was a wow. All of them loved global greening and the media loved them.

And there was another very important player—Senator Al Gore, Democrat of Tennessee—who arrived early, was popular with the media and was obviously making friends he could use later as President Bill Clinton's emissary. Rio was Gore's wheelhouse.

The *Economist* ran a cartoon showing Bush arriving at the Rio function. He's surrounded by heads of state from all these countries, smiling and applauding, while one of them is pasting a sign on his back that says "Kick me."

Bush made his remarks, sat through a session or two, with as low a profile as the leader of the world's most powerful nation can assume, and returned home. It wasn't terrible, but it was no fun in this last year of his presidency to have to run the multinational gauntlet.

By comparison, the current President Bush made a considerably smoother strategic move. Unveiling his epiphany on global warming, calling on his G8 economic brethren to join the U.S. in setting a target to cut carbon emissions in 2008, and doing this before he was put on defense—well, he did what we in the corporate sector have at times advised our executives to do when the cards are not stacked in the company's favor: take charge. A bold lead can sometimes defuse—or at least momentarily confuse—the opposition.

TRACKING CARBON[25]

As governments try to limit carbon emissions through emissions trading, regulators must track CO_2—where it comes from, where it goes— to verify transactions. Researchers have turned North America into a test lab. No other region is so thoroughly monitored; nor does any other area emit as much carbon dioxide—about 27 percent of the world's annual total. In his Wall Street Journal *column in December 2007, Robert Lee Holtz provided an overview of tracking mechanism under way in the U.S. This is an excerpt.*

Since the beginning of the Industrial Revolution 250 years ago, roughly 315 billion tons of carbon have been added to the air from the use of fossil fuels, land use changes and cement production, according the U.S. Department of Energy's online calculation of global CO_2 trends.

Even so, that's only about half of the total from human activities during that time. The rest was absorbed by the oceans, forests, grasslands, soil, peat and other natural carbon "sinks." The U.S. Environmental Protection Agency's National Greenhouse Gas Inventory presents estimates of U.S. greenhouse gas emissions and sinks.

The U.S. Climate Change Science Program recently released its first State of the Carbon Cycle Report, offering a scientific summary of the carbon cycle in North America, which releases much more CO_2 into the air than it naturally absorbs.

A new federal data analysis system called Carbon Tracker[26] offers the first systematic glimpse into how the continent of North America naturally recycles this critical greenhouse gas.

Scientists at the National Oceanic and Atmospheric Administration (NOAA), who developed the CarbonTracker system, reported their most recent findings in The Proceedings of the National Academy of Sciences.

---

[25] Source: Robert Lee Holtz, Science Journal, *Wall Street Journal*, December 28, 2007; http://online.wsj.com/article/SB119880485275254475.html.

[26] http://www.esrl.noaa.gov/gmd/ccgg/carbontracker/

"Rising concern about the environmental crisis is sweeping the nation's campuses with an intensity that may be on its way to eclipsing student discontent over the war in Vietnam ... a national day of observance of environmental problems ... is being planned for next spring...when a nationwide environmental 'teach-in' ... coordinated from the office of Senator Gaylord Nelson is planned...."

—*Gladwin Hill article in the* New York Times, *November 30, 1969, five months before the first Earth Day*

"'Green' means continually increased efficiencies in both operations and products, and we don't see any end in sight. It's not a choice between profitability and responsibility. They go hand in hand."

—*George David, chairman, United Technologies Corporation*[27]

---

[27]From UTC Web site, accessed June 2008.

Chapter Three

Is Your Company Still Going Green?
Get on the Road to Sustainability

Going green! What does it mean? Send your Web search engine after the phrase "going green" and you'll get lots of choices. You can catch an Oprah show segment in which it was revealed that the average person is responsible for emitting 94 pounds of carbon dioxide every day, and Oprah's "Going Green 101" guests had tips like planting four trees per family member to offset the family's monthly carbon footprint.

You can get "Going Green" columnist Bryan Walsh's insights and updates each week in *Time* magazine. Or you can actually get into the game and "submit your Going Green ideas" via an NBC site that billboards the phrase. And, if you're a company communicator, you won't want to miss a backward dive into the 2007 "Going Green" issue of *Fortune* that featured 10 companies said to have gone beyond what the law requires to become "green giants" operating their businesses with super environmental accountability.

So what does it matter if "going green" has become the snazzy fallback phrase for hurried (or lazy) newspaper headline writers and the electronic keyboarders who bang out the promotional news bites crawling across the bottom of your TV screen? Beyond the fact that I'm no doubt over-sensitized to a word-pair first used as the title of my 1992 book, and maybe just a little perturbed that the phrase has taken off without me, there are at least three more rational reasons to question the easy use—and, I will submit, probable abuse—of "going green" and its remarkable stickability.

Reason One: Context

It is awkward and generally misleading to use "green" to embrace the current and developing energy/carbon issues that have put environment back at the top of executives' sociopolitical agenda. "Green" began as a focus on nature. Its origins were in conservation, protecting wilds and woodlands, clean streams, clean air. As a cause, it grew from "ecology"; the word spray painted on the first Earth Day[25] on the promenade of the UN Building in New York, not far from the statue of the warrior turning his sword into a ploughshare. In 1970, this was an intriguing way to grab popular attention. Television coverage and newspaper photographs of the spray-painted word appeared worldwide. However, while "ecology!" might have become the code word for a cause—like "solidarity!" or "liberty!" posted for public attention—the study of how organisms interact with each other and their physical environment was overly complex as a rallying cry.

The rising sociopolitical environmentalists (not necessarily ecologists) and certainly the news channels needed a term easier to grasp and would settle on "green." The cause would need targets easier to grasp. Moving beyond the Rachel Carson cause of birds not singing, it would aim at familiar examples of waste, pollution, litter. Its foremost target would be corporate clean-up of products and emissions.

At the government regulatory level, the green agency was the U. S. Environmental Protection Agency, with some traditional federal green jurisdictions in the departments of interior and agriculture. Green was not a particular concern at the departments of energy or transportation, as it is now, with grants and joint research projects on subjects such as hybrid automotive technology aimed at environmental as well as energy outcomes.

Now, placing the original strictly environmental cause squarely on the sustainability road, it is energy—related to carbon, tied to climate change—that has spurred Congress, California and 15 follower states toward regulatory policy that produces sociopolitical results. Energy would be tilted into dominance in government environmental jurisdiction. EPA would be told by the Supreme Court to regulate for carbon constraint. California would push and sue EPA for authority to regulate automotive efficiency authority. Companies would be and are now pushed by rules and public opinion to adjust "green" to embrace energy conservation and efficiency, alternative energy sources, products and services. Green, as the sociopolitical issue born of ecology, now puts carbon constraint into business' previous accountability for care and kindness to people, animals and nature. Inside companies, going green effectively now means going carbon neutral, an entirely different proposition than original mandates. That is the core of advanced greening strategies in corporate executive suites.

Reason Two: Tense

You can also quarrel with external plugs about companies *"going"* green on the grounds of improper tense—present tense distorting the fact that American business has been *doing* green for decades. EPA has interacted with business on green policy and regulations for 30 years. Environmental management (often described more broadly as environment, health and safety management) is a mature corporate function, adjusting over the years to regulatory roll-outs, developing technologies, responding to markets.

Of course the tasks continue. Environmental management is the Aegean stable of social issues. But companies have grown green. Business has grown green because government required it, stakeholders wanted it, and it was the right thing to do—socially, politically if not (as it often was) financially.

Reason Three: There is a Better Term

If popular use of the term "going green" is misleading by stretching over carbon or anachronistic by putting it into present tense, should business communicators who know that language is the power punch in message effectiveness object seriously? Can we get the media off the "going green" mania? Is there a better way to say what's happening now and to take charge of the language for the future? I think the answers are no, no, no, and yes, but ...

"Sustainability" qualifies as the accurate name of the game, with *corporate sustainability* as the business derivative. Introduced as *sustainable development* at the United Nations Earth Summit in 1992, sustainability was linked with the American green movement in early observances of Earth Day. By definition perpetually focused on the future, sustainability is now positioned on a long public policy runway.

Corporate Sustainability Links Social and Economic Factors

Within companies, *corporate sustainability* is in effect the pragmatic successor to corporate social responsibility, linking social accountability to economic accountability. As an expansion of greening, it wraps around climate change, mobile and plant energy efficiency, fuel supply, and all the environmental/conservation aspects of greening in its various iterations, even forest and crop management. Old forests are not only to be saved for social, natural beauty and heritage reasons; they are part of the carbon-absorbing answer to global warming, the same rationale as the fervor to plant new spreads of trees, as the *Oprah* show advised its viewers and as many companies have either done directly or through carbon offset purchase. Crops became part of the global warming equation as an accompanying fervor pushed the planting and use of corn, soybeans and grasses in biofuels.

Adjustments in company communications follow the flow and define the new meaning of greening. In a subsequent chapter, we will examine company Web sites and see how green orientation has merged into corporate citizenship and environmental responsibility has expanded to embrace sustainability. An example of this amalgamation was the move in 2007 by Ford Motor Company to add a *sustainability officer* to its C-suite team. Ford as well as several other large companies have made it easy for stakeholders and the media to see sustainability as the current, larger picture by revamping their Internet portals and summarizing the ways in which various environmental and energy efficiency efforts, products and assets comprise the company's *sustainability platform*.

"Going Green" is Code Green

At this point, your author puts aside his personal, Bill Safire-type discomfort with mishandled language. Adjusting to one of the many things that I cannot change, I have found peace with "going green." A recent "going green" Google search turned up 105 million links to the phrase. There is a sizable entry in the do-it-yourself encyclopedia link entitled http://en.wikipedia.org/wiki/Going_green. My aging book with the original 1993 copyrighted title *Going Green* is still listed on Amazon.com.[28] Used copies sell for a pittance but it hangs in there and is running with the

---

[28]*Going Green: How to Communicate Your Company's Environmental Commitment*, Business One Irwin, Homewood, IL (subsequently acquired by McGraw Hill business books), ISBN 1-55623-945-9, with six topics listed in Library of Congress Cataloging-in-Publication Data: 1. Industrial management—environmental aspects—United States. 2. Public relations—corporations—United States. 3. Business enterprises—United States—Environmental aspects. 4. Green movement—United States. 5. Environmental protection—United States. 6. Social responsibility of business—United States. In addition to Amazon.com availability, the book can be ordered for reference and basic green guidelines through http://www.envirocomm.com.

Chapter Three Is Your Company Still Going Green?
Dealing with the Signs on the Road to Sustainability

younger crowd that believes or certainly hears that this company or that is "going green." So it may be achieving its own form of sustainability.

More important, the fact is that any external "green" description of a company's actions and assets put the firm on the positive side of the sociopolitical ledger. More than a halo, and certainly more than a self-anointed halo, "green" has become a strategic code word, a commonly understood indicator, signaling to stakeholders that the company cares about its impact on society. The company avoiding climate change risk and grasping climate change gold will be called "green" by investors, including some who were cool when the term just meant money spent without a return. The company flashing code green tends to improve its chances of receiving the stakeholder permission that pioneer business communicator Arthur W. Page[29] said is the key to business success. As Dan Esty and Andrew Winston said in their book[30] on environmental strategies, "green is gold." As the GE CEO said when launching its carbon-sensitive businesses, "green is green." And, perhaps topping the arguments to salute "green" as the widely accepted mark of American corporate distinction, there was the mid-decade observation by the *New York Times'* Thomas Friedman: "Green is the new red, white and blue."

The point is that corporate people need waste no time resisting implications from external messages that your firm is just now going green. It's usually a compliment and it can complement your plans and record of going sustainable.

As a one-time newspaper reporter with fearful respect for the copy desk chief whose decisions on my story drafts could make or break my day, I can further attest that we should not expect a linguistic leap in the news media from "green" to "sustainable" in describing what business is doing or not doing, if for no other reason than the fact that "going sustainable" is not as sticky a message as "going green" and is usually too long for a

---

[29] Page, called the "father of corporate public relations" in the biography by Noel L. Griese (*Arthur W. Page: Publisher, Public Relations Pioneer, Patriot*, Anvil Publishers, 2001) because of his contributions to communications philosophy and sound business practices, as AT&T's first public relations vice president (from 1927 to 1946) was himself directly engaged in the sociopolitical process of his day. He helped found the Center for the Study of Liberty at Harvard University, worked as a strategic communications adviser to World War II government agencies, and is credited for drafting the public statement of President Truman issued from the White House in 1945 on the first American war-time use of the atomic bomb.

[30] *Green to Gold: How Smart Companies Use Environmental Strategy to Innovate, Create Value, and Build Competitive Advantage* by Daniel C. Esty and Andrew S. Winston.

headline. An online article in 2007 about Hewlett-Packard's sustainability initiatives bore the headline "HP is Going Green(er)." Again, a few days later, a Sunday *New York Times* Art & Leisure section article about the surfer/singer/songwriter Jack Johnson becoming "a more intense environmentalist" was introduced with a teaser head: "Going Green(er)."

I'm heartened to think that some copy desks now realize that "going green" is not going far enough and we will see other small—positive if at times clumsy—steps in the direction of recognizing the expanding reality of sustainability. Meantime, companies are by their actions, strongly motivated by energy and climate change considerations, defining the terms, benefits and strategies of *corporate* sustainability.

GREENHOUSE GAS OVERVIEW[31]

Gases that trap heat in the atmosphere are called greenhouse gases (GHGs). Some greenhouse gases such as carbon dioxide occur naturally and are emitted to the atmosphere through natural processes and human activities. Other greenhouse gases (e.g., fluorinated gases) are created and emitted solely through human activities. EPA says these are the principal GHGs that enter the atmosphere because of human activities.

• Carbon Dioxide (CO_2): Enters the atmosphere through the burning of fossil fuels (oil, natural gas, and coal), solid waste, trees and wood products, and also as a result of other chemical reactions (e.g., manufacture of cement). Carbon dioxide is also removed from the atmosphere (or "sequestered") when it is absorbed by plants as part of the biological carbon cycle.

• Methane (CH_4): Emitted during the production and transport of coal, natural gas, and oil. Methane emissions also result from livestock and other agricultural practices and by the decay of organic waste in municipal solid waste landfills.

• Nitrous Oxide (N_2O): Emitted during agricultural and industrial activities, as well as during combustion of fossil fuels and solid waste.

• Fluorinated Gases: Hydrofluorocarbons, perfluorocarbons, and sulfur hexafluoride are synthetic, powerful greenhouse gases that are emitted from a variety of industrial processes. Fluorinated gases are sometimes used as substitutes for ozone-depleting substances (i.e., CFCs,[32] HCFCs, and halons). These gases are typically emitted in smaller quantities, but because they are potent greenhouse gases, they are sometimes referred to as High Global Warming Potential gases.

---

[31] Source: U. S. Environmental Protection Agency.

[32] CFC or chlorofluorocarbons, once commonly used in aerosol sprays and as refrigerants, were banned and U.S. production stopped as the result of the Montreal Protocol of 1987.

"Sustainability 360 takes in our entire company—our customer base, our supplier base, our associates, the products on our shelves, the communities we serve. And we believe every business can look at sustainability in this way. In fact, in light of current environmental trends, we believe they will, and soon."
—*Lee Scott, CEO, Wal-Mart*[33]

"There's not a sustainable company or sustainable product anywhere on earth yet, but we're working on it and making terrific progress."
—*Ray C. Anderson, CEO, Interface*[34]

---

[33]From Scott's keynote lecture at the Prince of Wales' Business and the Environment Program in London in 2007, announcing Wal-Mart's broad sustainability program.

[34]Anderson founded Interface Inc. in 1973 and built it into the world's largest producer of modular floor coverings. As explained in his book *Mid-Course Correction*, he committed his company to a path of zero emissions in 1994 and has become an outspoken advocate of sustainability. Quote is from an interview in *@issue*, published by Corporate Design Foundation, Spring 2008.

Chapter Four

Corporate Sustainability: Balancing the Tripartite Accountabilities

The word "sustainability" has become so overused that it has lost precision. While this book's scan of the various applications of sustainability will no doubt continue some of the confusion about what sustainability means, my intention is toward clarity when it comes to business strategies and what I call sustainable business communications. That is why for our purposes I have given "sustainability" a first name—*corporate*—and have defined it as what a company does to align or balance specific accountabilities that are pulled into focus by consideration of energy and climate change relevancies.

Think of corporate sustainability as the sign in the executive level board room, placed on a tripod. Think of the three legs supporting the message as social, economic, and political accountability. If you are CEO or communications chief in the C-suite, the question, as you rethink green and try to move to sustainability, is: *does the tripod wobble?* How do you keep the tripartite factors reasonably balanced? How do you avoid negative tipping points that shake up stakeholders and stir up critics? What lessons have been learned by company CEOs and chief communicators moving along the path of sociopolitical greening, preparing for the socioeconomic challenges ahead?

The private sector in the United States has adjusted remarkably to social and political demands beyond any required elsewhere in the world, accepting constantly expanding accountability for environmental impact and making extraordinary progress toward an ultimate dream goal: zero pollution from stationary and mobile sources. In the process, C-suite executives in a wide range of companies have made the environment part of their social contract with stakeholders.

Sustainable stakeholder communication is critical to this process, and far from easy. Proliferating issues, the burden of laws and regulations, and the supercharged voice of antagonists, joined by politicians, pumped by media, casting blame on business have produced that "over-communicated society" where, as Al Ries and Jack Trout once observed, very little communication actually takes place. To break out of isolation and be part of the answer in the over-communicated green dynamics, companies have learned to be very specific in aims and achievements and to focus foremost on those relied on for consent to conduct a successful business: the critical stakeholders. Company communicators realize a distinctive position must be created in the minds of its specific stakeholders, touting strengths credibly and honestly, and addressing its weaknesses and differentiating positively from the competition.

During the last two decades, broad goals of company mission statements, conveyed to stakeholders in annual environmental reports, many of which are now rebranded as sustainability reports, have come to include specific environmental and social responsibility, supplemented with pollution and waste-reduction targets.

INVITE INSPECTION

Technical progress is measured, communicated, and ready for stakeholder relevance. The "track us, don't trust us" executive message from Rio in 1992 is today's corporate sustainability guideline, the obvious basis for trust over the long term. In January 2008, an almost identical message appeared in the influential op-ed corner of the *New York Times* placed by General Motors. The ad invited the public to ask questions about GM's sustainability moves through a dedicated Web site. Companies now routinely open their operations to external inspection to validate reports of performance. Similarly, international environmental management standards such as ISO 14000 are widely followed, and cooperative industry standards such as Responsible Care in the chemical industry have put companies and their former critics into open and supportive relationships.

Early tactical communications mistakes such as the TV commercial showing seals applauding chemical emission reductions have helped corporate communicators understand that, if inclined to tout green, apply the red-face test. Let others brag; instead, practice a level of humility, invite inspection, and prove the company's commitment.

Chapter Four Corporate Sustainability:
Balancing the Tripartite Accountabilities

Challenges to Come

Third-party recognition and awards have always been easy to gin up among cohorts. What's changed is that company and hard-core environmental communicators have found public-interest synergy. Companies launch initiatives to reach clean-up goals urged by the environmental community and create measurable good-citizen records. An example with which I am most familiar was the move by Navistar's International Truck and Engine years before regulations required it to manufacture and deliver to school authorities in California, a no-smoke, no-smell diesel bus that won the approval of government and environmental advocates, and enabled the company to succeed in the school bus business.

Investors and directors are told the business case for environmental initiatives and plans for handling financial impact of compliance. Corporate communicators through web sites speak credibly 24/7 to stakeholders, reinforcing environmental/social responsibility missions, achievements and commitments, validated by articles, fact sheets, photos and links to third-party sites. Associations have formed along modern social fault lines. Partnerships within and across private and government sectors, and increasingly involving advocates who have fought industry, are the 21st century face of action.

Arthur W. Page, esteemed as a pioneer in solving the puzzles of compatible company-public relationships as he found them as the head of public relations at AT&T beginning in 1927, said a company operating in a democracy relies on consent of the public for its success. The company's operations, its sales, its profits, its *sustainability*, all rest ultimately on permission from direct stakeholders. Activists, media, and politicians get into the act when social compacts with stakeholders are unmet or unsatisfactory.

Undoubtedly, more sociopolitical challenges are ahead. Green issues will continue to enliven political campaigns. There is virtually no limit to what technology will discover and critics will uncover to challenge business-social dealings. Obvious current examples range from the macro, like global warming, to the micro, like nanotechnology, that will both solve some environmental problems and discover new levels of pollution that are as yet not measurable.

Business Uncertainty

We cannot now know how the current challenge of global warming will affect business in general and certainly any particular company. Uncertainties abound. Just as you think the consensus is forming around acceptable and available alternative fuels, biofuels are questioned. Plan to play in the carbon-emissions-trading market, and Europe, the player who started the game, starts questioning the original rules. Try to pick a price for future carbon emissions, so you can invest soundly, or try to offset carbon footprints by finding a trustworthy tree-planting operation, or try to figure out what your B2B green-supplier demands will be or how far out front your CEO should go and not have the rug pulled out from under him or her—well, like so much in corporate life, certainty feels like the impossible dream.

Fit to Financial Performance

My point is that American companies are increasingly, and I would say intractably, engaged in the sociopolitical process surrounding green issues; they are fitting accountabilities in this rumbling arena to successful financial performance.

Corporate sustainability means building a business that will last because its financial performance is not handicapped by social and political factors.

Calls from outside the business community, whether they are real demands or political/activist rhetoric, for greater social performance will ring increasingly hollow against the evidence of companies who are reasonably balancing social accountabilities with the other two legs of the tripartite accountabilities—economic and political. Supported and facilitated in the executive suite by its communications professionals, corporate management will need to sustain the ability to adapt to changes inherent in the current carbon war and to continually earn consent of public evaluators, starting with the specific company stakeholders.

Stakeholder engagement circles are expanding with public awareness that climate change is a problem and business is central to the answer. Wal-Mart, the world's largest retailer, and probably the most substantial *Corporate Greening 2.0* influence, looks for solutions with what could be the model for sustainable business communications: a company-wide emphasis on sustainability extending beyond Wal-Mart's direct environmental footprint to engage associates, suppliers, communities, and customers. The name of

Chapter Four Corporate Sustainability:
Balancing the Tripartite Accountabilities

the program—"Sustainability 360"—suggests the interactive, synergistic review process familiar to companies and proven in stakeholder engagement managed by corporate communications and human resources.

BREAK THE C-SUITE SECRECY

Commentary: August 2, 2007—"Quick, boss, apologize!" Are these the first words out of the mouth of the chief communications officer when the CEO owns up to personal or professional transgression? I don't think so.

More likely, the CCO is thinking: *Maybe now they'll listen to me. You can't do this stuff. You've got to play it straight with the stakeholders.* So he or she pulls up to the management table and says, "Okay, what do we know and when did we know it? Never mind why. Just … what's happened that somebody thought we could keep secret?"

John Mackey of Whole Foods brings this to mind. His double life—by day, running a business; by night, using his wife's pseudonym and chatting with Web strangers—ends up as laughable news stories, with the biggest laugh a heartfelt apology for his "error in judgment." More CCOs can probe and preach transparency with the support of bosses who understand the reality of openness.

That's not what this column is about. Frankly, I find I don't care much about Mackey or any of the other chief executives who have in the past months and years expressed contrition for one thing or another. I do care very much about chief communications officers (having been one and knowing many), and what this cleaning-up-after-the-boss deal means, and that's what I believe needs their serious attention.

My interest is in how the CCO moves up in management to help break the chain of missteps and apologies, which I see as a huge hazard in the CCO's critical wheelhouse of stakeholder trust.

To influence change, CCOs will need to probe a little deeper and preach a little harder in the C-suite. Detachment from information, the incremental grist that precedes stakeholder communication, is professionally fatal to the chief communicator. Somehow, the CCO has to know constantly what's going on in HR, legal, operations, and financial that has any potential to become part of the dialogue between the company and its workers, owners, business partners, government, communities, and other public stakeholders.

Probing—digging for intelligence and sifting for potential danger—is now, more than ever, job one. Much of the corporate apologia syndrome is rooted in the belief that secrets can be kept. That's a dangerous throwback. It distorts and clogs the flow of information that, whether it's good news or bad, shakes the trust/permission deal with stakeholders.

As guardian of this vital deal, the CCO needs the reputation in the suite as the key good guy. He or she is the person with the very practical purpose of detecting information flow problems, the one who asks questions about what's going on in operations, in legal, in HR—in all the thickets where stakeholder trouble grows—and thereby protects his or her fellow suite dwellers from the agonies of future discovery.

This ties into CCO job two—the how of the role as transparency officer; that of conveying with some authority and passion the evidence and the ways in which the corporate communications world has changed.

In this transparent world, not only does everybody eventually know everything; because of the Internet, everybody can communicate about everything, all the time and forever. Like preachers, the C-suite communications officer may need to get the attention of his or her suite mates by invoking the wages of attempted secrecy. Words are Web everlasting, accessible to everybody, Googleable, bloggable ...

CCOs' role as guardians of trust will become easier. Cowboy CEOs—whose mantra was "never complain, never explain"—are already fading away. Steady veterans in the Warren Buffett model understand that investors and other stakeholders are reassured by open kimonos and plain explanation.

Younger chiefs, rising in the Sarbanes-Oxley searchlight, generally know they're the accountability officers, accountable to boards, investors, and the stakeholders who are connected and aware as never before. Unlike the risk-happy, after-hours CEO blogger, most have healthy risk-avoidance instincts. Consequentially, more and more CCOs can probe and preach on transparency with the support of bosses who understand the reality of openness and are determined to minimize having to say they're sorry.

These conditions favor chief communications officers becoming substantially stronger players in their C-suites. Heroes even.

GREENEST CITIES

Which American city leads all others in taking action in the war on carbon? According to the Internet search engine Yahoo, it's the birthplace of Kool-Aid and the 2007 winner of Yahoo's "Be a Better Planet: Greenest City in America Challenge": Hastings, Neb. Here, from Yahoo.com is Yahoo's "Top 10 Greenest Cities List for 2007":

1. Hastings, NE
2. Pelzer, SC
3. San Carlos, CA
4. Mill Valley, CA
5. Topeka, KS
6. Dover, DE
7. Spring, TX
8. Lawrence, KS
9. Walnut Creek, CA
10. Fairfax, VA

Hastings Mayor Matt Rossen told Yahoo its $250,000 prize will help local citizens follow through with eco-friendly commitments like recycling, taking public transportation, using resuable shopping bags and changing light bulbs.

Source: Yahoo.com

"Entrepreneurial capitalism is the strongest force in the world. This may sound surprising coming from an environmentalist, but we've seen profit and innovation work to solve acid rain and other problems. Properly channeled, these two historic drivers of human enterprise can save our planet from catastrophe. The Industrial Revolution started global warming. A new clean-energy revolution can stop it."
—***Fred Krupp, President, Environmental Defense Fund***[35]

"My view of making change happen is to work on persuading opposition that their strategy has some flaws. Part of that is convincing groups like the fossil fuel industry that just trying to say 'no' to global warming … is doomed to failure."
—***David Hawkins, Climate Center Director, Natural Resources Defense Council***[36]

---

[35]Fred Krupp, Environmental Defense Fund president, in a promotional interview for his book, *Earth: The Sequel: The Race to Reinvent Energy and Stop Global Warming*, 2008. Written with journalist Miriam Horn, the book explores signs that the $6 trillion energy sector is changing to embrace green energy sources.

[36]Interview by Christa Marshall in *ClimateWire*, April 8, 2008; http://www.eenews.net/climatewire.

Chapter Five

Seat at the C-Suite Table: Putting Climate Change into Competitive Business Strategies

An environmental think tank, World Resources Institute, with help from the investment research staff at Sustainable Asset Management, has calculated the "climate competitiveness" of selected global companies by using two variables: positioning against risks, and preparedness to seize opportunities. By simply zeroing in on these two factors, WRI's Jonathan Lash told company management in a *Harvard Business Review* article, "you are likely to uncover ideas on how to move to a position of competitive advantage."[37]

Executives at the C-suite strategy table considering revision to previous strategies as the war on carbon escalates can begin with an organizing principle: *To sustain our business success and to beat our competitors, we must know and successfully adjust to climate change dynamics.* Corporate greening strategies require input and interaction across the scope of the company's business. This means bringing to the table the firm's business unit leaders to provide perspective on business, product, sales, and service-impact probabilities, and to learn what they must know as they plot revenue and profit targets. The chief financial officer will need to help set the climate-change perspective for financial and accounting performance. Government relations will provide an overview and apparent direction of both the relevant pressure likely to come from federal, state, and local governments, as well as any incentives in the form of government grants or tax relief that can affect the company's business decisions. Environmental management (the company officer specialized in the field of environment, health, safety and energy) will be asked for insights on conditions, costs, and prospects for carbon constraint in operations and product stewardship.

---

[37]"Competitive Advantage on a Warming Planet," by Jonathan Lash and Fred Wellington, *Harvard Business Review*, March 2007; also see http://harvardbusinessonline.com.

Legal and insurance risks (or if the company is in the insurance business, the prospect of coverage products) must be factored into decision options. Human resources, security, IT, and any other aspect of the business touched by carbon dependency and emissions—and others as determined by the CEO or the designated climate-change or sustainability officer—will need representation as collaborators toward new patterns of corporate sustainability.

CCO Role

The chief communications officer will facilitate the exploration of corporate sustainability by helping others in the C-suite understand stakeholder and corporate reputation risks and opportunities tied to climate change. He or she will know the *social capital* (a concept explored usefully by Francis Fukuyama in his book *Trust*[38]) that is valued by each stakeholder group: jobs and work conditions for employees, economic return and performance evaluations for investors, orders for suppliers, desired goods at the right price for customers—and how change impacts the trust/permission deal. The CCO will take responsibility to assure executive-level information flow. Each of the other executives will value being adequately informed about developments that affect the sustainability deal for good or bad—information on what's happening in other companies, on resources available to think through strategic problems, on events or channels to communicate. In many companies, the CCO will also provide the nudge factor—making the rounds of informal conversation, gathering information and data, connecting the dots and bringing facts and inferences into specific relevance for the other members of the sustainability team—to enable team play.

Questions at the Strategy Table

Corporate Greening 2.0—recalibration of green strategies and carbon-constrained business plans—will center on questions such as these:
1. *Assessing the Risks:* What are the dimensions and how would we manage carbon-constraint risks to the company—regulatory,

---

[38]Fukuyama's book, *Trust: The Social Virtues and the Creation of Prosperity* (New York: Free Press,1995), is a fascinating exploration of social trust, comparing high-trust and low-trust societies. Social capital is used to describe what people in a particular culture or community consider valuable.

Chapter Five Seat at the C-Suite Table:
Putting Climate Change into Competitive Business Strategies

product and technology, supply chain, insurance, litigation, reputation, and physical risk?

2. *Managing Carbon Constraint:* Who takes management accountability inside the company? Do we assume the current environmental management team can handle the carbon-footprint management, reporting, and other changes? Does our ISO or other environmental standards program fit the energy management plan?

3. *What Stakeholders Expect:* How will government decisions affect us and our stakeholder relations? What will company stakeholders expect of our company in view of carbon-constraint changes? What is the potential for change in our stakeholder base as the result of product, production, and other management decisions?

4. *Localized Global Effects:* Do we need to examine the company's vulnerability to, and possible opportunity in, *local* climate change dynamics, local regulations, civic action, supply, and consumer choice?

5. *The Science:* What are the implications of climate change and energy science contexts on our business plans? Do we need an independent study or survey of available scientific information relating to physical, geographic effects on our operations, locations, and business lines?

6. *Caps, Trades, Offsets:* What is the relevance of cap-and-trade schemes to our business? What are options we can consider in emissions trading and offsets? What risk/returns are attached to each option? What can we learn from the European experience and that of other companies? What are the trading exchange options, how are they managed, and what companies are engaged?

7. *Business Coalitions:* What are the reasons, options, and timing for the company to join (or organize) coalitions—legislative, regulatory, community, business partner, marketing, public relations objectives? What are the collaborations we should consider—who else is in them, what level executives, and what are the results so far?

8. *Government Incentives:* What are the innovation or development opportunities—such as lower-carbon technologies—encouraged, or likely to be, by government tax relief, grants, purchase or other incentives?

9. *Federal and State Politics:* What are our political contexts—how Congress responds to the carbon war, the role of the states, local/municipal government involvement? How are our political relationships in these contexts?
10. *Regulatory Outlook:* What is the regulatory outlook for Federal and state agency rules implementing carbon constraint rules? How will this affect compliance of our operations, products, services, and locations? What can we learn from current regulatory relationships?
11. *Reporting Carbon Reduction:* What are we reporting now and how is this likely to change? Where are we on required and voluntary options in providing "carbon disclosure" information? What are our options regarding potential government, NGO, and private organizational reporting?
12. *Communications:* How does management communicate? What are the benchmarks for communication relevant to our field? What can we learn from the way other companies are communicating: management messages, use of the Internet, and terminology used to engage stakeholders? Do we need, and if so, how will we effectively manage interactions online?

STAKEHOLDER FOCUS: DRUCKER'S 3 QUESTIONS

The common thread running through all of this is knowledge of stakeholder thinking. Company executives forming climate change and sustainability strategies can use a variation of the three questions that management guru Peter Drucker used to begin his counseling with client executives. He would ask: *Who are your customers? What do they need, want, or expect? And, therefore, what do you do?* The point is to shift the executive mindset to customers. This, said Drucker, is essential because management's first job is customer creation. Forming sustainable strategies requires sustainable relationships with key stakeholders. Employees, customers, investors, suppliers, geographic neighbors, global business partners—the entire range of the firm's business universe—will require attention. For some companies, attention to the possibility of new sets of stakeholders will be in order. Collaborations of a different type may be the key to economic, social, and political success.

Chapter Five Seat at the C-Suite Table:
Putting Climate Change into Competitive Business Strategies

Green Collaborators in TXU

The TXU utility deal of 2007 is an example. The coal-powered Texas utility was not acquired by its Wall Street financiers until it settled environmental issues with help from a green community that in the Greening 1.0 period had adamantly opposed the continued existence of coal power. The settlement established the reality that previous non-financial issues—the sociopolitical issue of climate change being the specific example in this case—are now widely accepted as having financial materiality. It allows the coal company to stay in the energy game, providing it can work out the moves in a complex but achievable future scenario with green community collaborators.

Three of the key players working at the table with industry and investor representatives were Fred Krupp of Environmental Defense Fund; David Hawkins of the Natural Resources Defense Council; and William Reilly, former U. S. Environmental Protection Agency administrator, now with an investment group.

Krupp, EDF president, calls himself a "market environmentalist." In addition to his and EDF's central role in the $45 billion buyout of TXU, he was also instrumental in the launch of the U.S. Climate Action Partnership, helping major corporations move towards support of a nationwide cap on carbon, and he works closely with Wal-Mart's leaders to influence market and supplier greening worldwide.

An intelligent, reasonable and careful partner with business people in pro-green initiatives, Krupp is no pushover. His partnering hand with companies began as a fist of protest, raised against the environmental impact of McDonald's in the 1980s. This ended with a handshake agreement that McDonald's would move away from non-biodegradable cups and sandwich containers, and work with EDF on green public strategies. McDonald's and EDF in 2008 celebrated the 10th anniversary of their green alliance. Krupp and McDonald's CEO Jack Greenberg jointly announced the elimination of 150,000 tons of packaging eliminated, millions of kilowatt hours saved, and billions of dollars spent on recycled goods during the collaborative decade. This is a clear indication of the way that Krupp has entered into the business community with effective results that have inspired other NGOs as well as some business people to reach toward green partnering. Krupp now advocates the potential for the energy sector to transform itself into a profitable force for carbon constraint through new technologies. His book, *Earth: The Sequel*, emphasizes the potential for capitalists to solve climate change issues, noting that inventors and entrepreneurs are already "on the frontiers of energy" with such innovations as sending smokestack gases from coal-fired power plants to pools of algae that convert to biofuels. Although Krupp is not a member of the new advisory board of the reborn TXU (the position is held by EDF's Texas leader), he will be watching the Texas operation for opportunities to break through conventional technology barriers.[39]

CLEAN AIR LAW KING AS COLLABORATOR

David Hawkins is another tough veteran figure in American environmentalism. In the 1970s, as Greening 1.0 took off with the largest wave of environmental laws in the nation's history (three times as many laws were enacted in this period than in the preceding two centuries of the Republic), Hawkins was a dedicated staff attorney with the Natural Resources Defense Council in Washington, DC. With others who helped make NRDC the fighter pilot in the war on American pollution, Hawkins initiated the group's clean air project and led its force to reshape the Clean Air Act (known originally as the Air Pollution Control Act) and pressure the Environmental Protection Agency toward stiffer provisions. In 1977, he was nominated as EPA Assistant Administrator for Air and Waste Management by President Carter and served in that post until Carter's re-election bid failed in 1980.[40]

With Ronald Reagan's presidency, Hawkins returned to NRDC to engage in the interplay of government, business, and NGOs. He was a major figure in a green community which was now energized to resist what they felt was a retreat from the level of intensity on environmental matters achieved by the Democrats and their combined forces. With the century-turn emphasis on energy and climate change providing a fresh

---

[39]With offices worldwide, from Boston to Beijing, and in Bentonville, Arkansas, as a neighbor to Wal-Mart's corporate offices, EDF has positioned itself as the leading nonprofit developer of market-based environmental solutions. See http://edf.org.

[40]Hawkins and fellow NRDC attorney Dick Ayers were like the "twin towers" of the green movement in the 1970s, pushing the movement forward towards funding and influence. Frank O'Donnell, president of Clean Air Watch, told *ClimateWire* reporter Christa Marshall, "It was . . . a hard core of guys coming out of law school who wanted to be do-gooders." (April 8, 2008).

Chapter Five Seat at the C-Suite Table:
Putting Climate Change into Competitive Business Strategies

wind of NGO opportunity, Hawkins stepped forward to lead his group's charge, becoming director of the NRDC Climate Center to advance political attention—and legislative policies.

Contrasting with the Greening 1.0 days of us-vs.-them, Hawkins saw climate change as a game change. Common interests could bring former corporate combatants to the same side of the table with green activists. In 2007—exactly 10 years after Hawkins and NRDC first set its sights on climate change legislation—Hawkins, once called "the king of the Clean Air Act," sensed that he was moving towards his biggest victory yet. In preparation for congressional action on global warming that seemed inevitable to some corporate executives, Hawkins helped to form the U.S. Climate Action Partnership with executives from DuPont and Duke Energy. His push: develop plausible principles that can widen the carbon control game to as many players as possible, effectively turning combatants into collaborators. Together, they would ask government to regulate carbon emissions and to do it in a way that would provide companies some social, economic, and political certainty as well as some latitude in achieving results.

REILLY AS GAP BRIDGER

Bill Reilly is credited with bringing the potential collaborators to the Texas utility's negotiating table and moderating the ultimate agreement. Reilly's past suited him to the role. Both as investor and environmentalist, he has a record of bridging gaps between corporate and NGO communities. He has headed green groups, he's served at the highest levels of government, and he has been a director of DuPont since 1993. Evidence of his crowning Texas achievement is the fact that when the deal went down, Reilly was chosen to chair the advisory group that would counsel the utility's new owners.

Reilly's entry in government service was far less combative than Hawkin's. Yale, Harvard, and Columbia educated, with a brief tour of Army duty in Europe, Reilly was working on urban beautification (a forerunner of environmental justice issues) at Urban America, Inc., when he was tapped in 1970 to become a senior staff member of the President's Council on Environmental Quality. CEQ Chief Russell Train, whose background included serving as head of The Conservation Foundation (TCF), was subsequently appointed by Richard Nixon as the second EPA Administrator. In 1972, Reilly left the government post at CEQ to follow

Train's footsteps as the leader of TCF. When TCF merged with World Wildlife Fund in 1985, Reilly became president of WWF. He headed this global organization until 1989 when President George H.W. Bush gave him the chance to come back into the federal bureucracy—this time as EPA administrator. At the 1992 U.N. conference, where the national policy concept and the corporate initiative on sustainable development were launched, the American environmental administrator, William K. Reilly, was the nation's most prominent figure. Reilly came to Rio several days in advance of the President, taking the brunt of media interviews, activist entreaties, and, between press conferences, briefings and staff preparations for a presidential arrival, took time to meet with the CEOs planning the business sideshow.

After leaving EPA during the final days of 1992, Reilly returned to World Wildlife Fund.

He is now president and chief executive officer of Aqua International Partners, an investment group that finances the purification of water and wastewater in developing countries, and invests in projects and companies that serve the water sector. Aqua International is sponsored by the U.S. Overseas Private Investment Corporation and is part of the Texas Pacific Group, an investment partnership based in Fort Worth and San Francisco, which invests in environmental, airline, apparel, health, wine, technology, and other companies. With the TXU deal confirmed, there was little surprise in the announcement that the first chairman of the sustainability advisory board, with representatives from business, environmental, and government sectors all around the same table, would be Bill Reilly, the gap bridger.

COLLABORATION CONSIDERATION

The economics of this story are not over. The company now known as Energy Future Holdings[41] faces challenges. The impact of government

---

[41]TXU Corporation was acquired in October 2007 by Kohlberg Kravis Roberts & Co. and Fort Worth-based TPG Capital. At $45 billion, it was the biggest leveraged buyout in corporate history, and became Energy Future Holdings. Charter members of the EFH sustainable energy advisory board, serving with William Reilly as chairman, were Reginald Gates, president and chief operating officer of the Dallas Black Chamber of Commerce; Karen Johnson, president and chief executive of United Ways of Texas; Sam Jones, former president and CEO of the Electric Reliability Council of Texas; Jim Marston, regional director of Environmental Defense Fund's Texas Office; Steven R. Specker, president and CEO of the Electric Power Research Institute; Texas Secretary of State, Phil Wilson; Ralph Cavanagh, co-director of Natural Resources Defense Council's Energy Program; and a representative of the Texas Association of Manufacturers, http://www.energyfutureholdings.com.

Chapter Five Seat at the C-Suite Table:
Putting Climate Change into Competitive Business Strategies

rules such as cap-and-trade is yet to be determined. However, it is a first, historically different from outcomes in Greening 1.0 and instructive in the way in which Greening 2.0 will unfold.

The point here is that collaboration strategies can be an important factor in preserving a company's sustainability options. Corporate sustainability is the hopeful process of aligning social, economic, and political accountabilities. While the process of linking with others will involve compromise, it can also remove risks and create new stakeholders who bet on and try to help achieve financial success even under greatly altered social and political conditions. Companies rethinking their failed or threatened green strategies in the light of government controls and public opinion related to climate change will do well to re-examine stakeholder bases, set aside negatives and desperation, and consider the possible benefit of collaboration.

Model Collaborator: IBM

IBM's extensive external collaboration and participation with government and nongovernmental organizations in energy and climate efforts, often being the first, or among the first, to support voluntary initiatives, and to set, meet or exceed aggressive targets. Here, from the company's Web site, is IBM's record of engagement with others on energy and climate change:

- 1992—Charter member of U.S. EPA's ENERGY STAR® Computers Program
- 2000—Charter member of World Wildlife Fund's Climate Savers Program
- 2000—First IT company to join Pew Center on Global Climate Change's Business Environmental Leadership Council
- 2000—Charter member of World Resources Institute's Green Power Market Development Group
- 2002—Charter member of U.S. EPA's Climate Leaders program
- 2003—Charter member of Chicago Climate Exchange®
- 2006—Qualified for and joined U.S. EPA's Green Power Partnership
- 2007—Founding member of The Green Grid℠

Another distinguishing aspect of IBM's energy and climate protection programs is the company's voluntary disclosure and transparency. IBM was one of three manufacturers to begin reporting its greenhouse gas emissions under the U.S. Department of Energy's Voluntary Greenhouse Gas Emissions (1605b) Reporting in 1995, a practice that continues. IBM has responded to the Carbon Disclosure Project since its inception in 2002. In addition, under the Chicago Climate Exchange, IBM's direct and indirect CO_2 emissions from its North American operations are verified by the National Association of Securities Dealers.

Source: IBM.com

COLLABORATING WITH GOVERNMENT:

Climate Business Partnership Programs

U. S. Environmental Protection Agency is an active facilitator for company climate change and sustainability strategies. EPA partnership programs offer technical assistance, financial and environmental analysis tools, training, seminars, networking, guidebooks and environmental performance benchmarking. In 2008, EPA and Business for Social Responsibility (BSR) released *A Business Guide to U.S. EPA Climate Partnership Programs*, a one-stop business resource for all of EPA's 35 climate-related partnership programs, which have more than 13,000 participating firms and other organizations. BSR said the guide minimizes the legwork necessary for companies to understand each program and identify a short list of programs that could work for a particular company. The EPA guide includes profiles of the agency's climate programs, organized by business value, services offered, program descriptions and applicable industries. It outlines how companies are investing in energy efficiency, clean energy supply and other climate-friendly technologies to realize a range of business benefits, including benchmarking and public recognition.

Sources: The business guide is available on BSR's Web site at http://www.bsr.org/reports/biz_guide_to_epa_climate_partnerships_508.pdf.

For more information about EPA's partnership programs, visit http://www.epa.gov/partners or http://yosemite.epa.gov/gw/StatePolicyActions.nsf/webpages/VoluntaryPartnershipPrograms.html.

Four federal programs aimed specifically at electronics are described at http://www.federalelectronicschallenge.org/resources/docs/other_partnerships.ppt

"Sustainability is part of everything we do. If you are in marketing, it's part of how you talk to customers about our products. In the financial sector and the business development sector, you are looking at how sustainability is going to be a growth platform. If you are in the manufacturing sector, you are trying to be more efficient in use of materials and to use renewables. It's part of everyone's job."
—Erin Meezan, Director of Sustainable Development, Interface[42]

"Most of my work is in operationalizing…three goals: developing and marketing $20 billion in… sustainable innovational products, improving the environmental footprint in operations, and the social sustainability area of increasing the [company's reach to] children in need…My job is to ensure there are principles across the company and that we abide by the principles."
—Len Sauers, Procter & Gamble[43]

---

[42]Interface is the carpeting manufacturer whose founder, Ray Anderson, was an early advocate of sustainability, after reading *The Ecology of Commerce* by Paul Hawken in 1993. Meezen, an environmental attorney, joined the company in 2004 to manage a high-level, global sustainability council that brings together key leaders from throughout the company. The firm is pledged to zero life-cycle impact of its products by 2020.

[43]Sauers, vice president, global sustainability, at P&G, quoted in the *Cincinnati Business Courier* in 2008, said the world's largest consumer products company would cut its carbon footprint by 40 percent by 2021.

Chapter Six

Changing Orientation: Sustainable Course for Corporate Communicators

Corporate sustainability is not a 21st-century idea. That is a good thing because it can be brought into the climate-change, carbon-constrained business challenges with an encouraging history. As green and energy issues come into focus in C-suite agendas, it is heartening to remember that the concept of coexisting social, economic, and political factors as a successful business strategy has been tested and proven workable.

At the world's first summit of nation leaders determined to save the planet, more than a decade ago in Rio de Janeiro, I was part of the sideshow. While the formal United Nations Earth Summit sessions tossed around calls for environmental treaties and the big idea of "sustainable development," a group of multinational companies conducted their own conference and I was invited to be on a panel to talk about ways that this would impact the business community.

While the exact meaning of sustainability when tied to the various national goals was not then clear—the general idea was not doing any economic activity nor causing any environmental impact that would rob future generations of their economic and environmental options—it was clear to us huddling in a corner of the U.N. tent that business was about to be pulled into a major task of trying to align environmental and economic goals for the long term.

Facing this challenge, and in fact showing remarkable initiative, a group of business executives at Rio announced plans to collaborate through a new group—the World Business Council on Sustainable Development—and to start providing the public with much more information on company green performance, including operational emission metrics.

QUALITY Model

Corporate communicators in our group discussed ways to take this management idea and make it relevant to company stakeholders. My contribution was a seven-step model for stakeholder engagement developed and tested by my EnviroComm team, known as our QUALITY model:

Quantify the stakeholder universe (count and sort customers, investors, and other stakeholders into accurate research categories);

Understand the needs and interests of each group; dig deeper by

Asking a standard set of questions and

Listening to volunteered input;

Interpret what you learn, relevant to what you need;

Take charge by coming up with alignment strategies, making your plans fit with their needs (starting with information plans and needs); and drive towards

Yes!

Reaching "yes"—the approval of company actions by stakeholders—is the endgame of virtually all corporate communications strategies. As Arthur Page counseled, no company success is achievable without it. Or to put it another way—which will reveal my inspiration—the goal is a mutually satisfactory outcome to problems of disagreement, as Harvard's famous negotiating process[44] instructs, captured in Roger Fisher and Bill Ury's highly recommended book, *Getting to Yes*.

My message at Rio was that a process such as this is needed to keep business communications open for future deals—essentially, preserving future options, the goal of sustainable development policies being discussed by the national leaders in the U.N. conclave next door. The output of the QUALITY process was dubbed *sustainable communications* and became the basis of my consulting practice, working with and learning from companies going green.

---

[44]Fisher and Ury's book *Getting to Yes: Negotiating Agreement Without Giving In* (now in a 1994 published edition) came out of the Harvard project in the 1980s. They counseled that the successful negotiator invents multiple options that ultimately yield mutual benefits. The thrust is to keep the game open—not to let it close with a "no" that can be moved toward a "yes," or at least to let each negotiator be aware that a "no" is not easily accepted as final. I benefited greatly from participation in two of the sessions held at Harvard, applying the principles in our EnviroComm practice. See *Going Green, Choose Your Green Partners*, page 190.

Chapter Six Changing Orientation:
Sustainable Course for Corporate Communicators

So, since at least 1992, corporate sustainability has been a factor in the business communications alignment with stakeholders on environmental and energy matters. But it was not mainstreamed. Despite its availability as a talking point, sustainability has been slightly tangential among C-suite topics: not as Greening 1.0 hard-core as cleaning up pollution to government compliance levels, nor as satisfyingly softer-side as a green social cause, nor as critical to reputation as financial performance.

Corporate Sustainability Mainstreamed

That has changed. Drafted for work as C-suites joined the world war on carbon, corporate sustainability is proving its value. The successful communications effort of Al Gore, popular recognitions of climate change at the 2007 Oscars and awarding the Nobel Prize to Gore and climate-change scientists, the personal drive of California Governor Arnold Schwarzenegger to put global warming into political play with his state in the lead, the 2007 return of Democratic control of Congress, the jolt of energy prices, felt at the gas pump level—these persistent and combining factors have tipped sustainability into practical business reality.

Public fear, elevated anxiety, professional risk analysis, competitive moves in the market place, high energy prices, and a strange new, game-changing factor—carbon dioxide, the exhaled emission, now deemed by government as a pollutant to be regulated—these new dynamics have combined to freeze basic environmental management at Greening 1.0 and to point companies toward the next level of economic, social, and political success: to *sustain* the business with both less pollution and less carbon.

This is Greening 2.0, the level of corporate sustainability that embraces new dynamics and plans for continued success. For some companies, like the coal-power utility on the front line of attack in the war on carbon, this means new forms of collaboration with environmental activism. For almost all companies, this means a plan to execute the vision of economic competitive advantage, enabling social harmony and political outcomes in which the company participates.

Sustainability Plan Guideposts

So now, joining peers at the climate change and sustainability strategy table, corporate communicators need to engage in a process along these lines:

- *Help form the business case.* Start with ties to financial performance. Chart the course for sustainability that's relevant to the company's strengths. This is a long-term program that will need dedicated resources. Top management and business operations buy-in and encouragement are critical to success. The case for sustainable communication is part of the business case.
- *Tie it to governance.* Since top management must get board endorsement and involvement, help connect the plan for sustainable economic performance to the sociopolitical factors. This can be a substantial corporate governance issue. In many companies, board members have responded to investor activist inquiries about corporate accountability for risks and opportunities related to climate change.
- *Internal alignment.* Internal communication has to achieve workforce involvement and operational accountability. This may be tantamount to a culture change. The flow of information within the company will need to assure that the whole company is aimed at sustainable performance. The CCO must help enunciate a clear definition of corporate sustainability, stated in terms relevant to the company's mission, appealing to its stakeholders and consistent with its values.
- *Stakeholder activism.* Each stakeholder group—investors, employees, customers, suppliers, government—needs to understand, and in the best case be engaged in, the company's strategies. Investor and government relations interactions are particularly important, because the opinions of stakeholders in these arenas can make or break management's resolve.
- *Reach out to collaborators.* Explore the options for partnerships with interest groups, government organizations, customers, and retailers. Understand what their needs are and whether your company can work collaboratively toward specific sustainability goals. Research or benchmark each potential collaborator's climate change commitments. And remember, the potential

collaborators, including a growing number of green former antagonists, may already be looking your way.
- *Think ahead.* Don't let the short-term mentality limit your future options. Prepare to succeed with the Millennium Generation of consumers and other stakeholders, recognizing that any company's sustainability is tied to reasonable expectation of future stakeholder benefits. Some companies are getting involved in school programs and public education about sustainability.
- *Transparency.* Realize that your company will be evaluated, with or without your help. Take charge. Accountability must be documented, and you'll need to show your homework to outsiders. Prepare to report progress (or lack of progress) in achieving metrics—such as carbon footprint reductions. Anticipate how your company will back up its promises, and get third party evaluations of processes and possibly its products.
- *Maintenance.* Create and use systems for internal and external feedback for continuous improvement in the sustainability program, keeping it relevant to the business and all stakeholders. Work with product stewardship as a sustainability mechanism.

WHEELHOUSE OF CORPORATE COMMUNICATORS

Stakeholder belief in the company's green commitments is vital, as the 2007 consumer environmental survey done by Carol Cone[45] and her associates in Boston documents. In this poll, 91 percent of respondents said they have a more positive overall impression of the company that is environmentally responsible, and as many as 81 percent of those surveyed indicated they would consider switching to another company's products or services if its corporate responsibility practices are not clearly positive.

The practical business value of corporate sustainability has come a long way from its roots as sustainable development in the 1992 Earth Summit. Climate change has clarified its meaning, raising its prominence in the C-suite as strategy for tackling sociopolitical issues. The response to global warming economics, social demands,

---

[45]Carol Cone is nationally recognized for her work in the cause branding and strategic philanthropy arenas. As chairman of Cone, Inc., she has embraced a steadfast commitment to building substantive and sustainable partnerships between companies and social issues for more than 25 years. Overall, Cone's signature cause programs have raised more than $500 million for various social causes.

and political influence combine to shake traditional social contracts. Corporate communicators can make themselves most useful as an advocate for connecting the company's climate-change geopolitics to the interests and activism of its various stakeholders. Because of his or her unique position as stakeholder surveyor—constantly scanning, connecting dots, and interpreting stakeholder input—and as architect of two-way synergizing information flow, the chief communications officer is positioned to understand and to play a lead role in engaging these stakeholders.

This uncommon change in business-critical dynamics may well provide a signal to a company's top management that it is time to review, refresh, or quite possibly renegotiate the trust/permission deal that every company has with its stakeholders, the critical arrangement that is heavily influenced by the content, flow, and timing of information—both outside and inside the company.

Corporate sustainability is a collaborative, cross-functional, mutually supportive management function that puts new kinds of information into decisions and relies heavily on sustainable communications. That places it into the wheelhouse of corporate communicators, whose first challenge may well be to help corporate people who still confuse it with previous silo programs of corporate citizenship and social responsibility.

Chapter Six Changing Orientation:
Sustainable Course for Corporate Communicators

COMMUNICATION STRATEGIES IN A SUPERGREEN OSCAR WEEK

Commentary: March 2, 2007—What lessons could C-suite PR chiefs glean from the greenest Oscar week ever?

While Hollywood glitterati at the Academy Awards, as well as pragmatic politicians among the TV audience millions, were swaying to the Al Gore rhythm, two more events fed a stream of strategies, reminders and a couple of benchmarks that corporate communicators could appreciate:

> *Don't give up on the strength of the executive or the message. . . . Remember the power of repetition, not to mention the PowerPoint. . . . Gain trust so you can manage to his advantage. . . . Grace works. . . . Set the agenda. . . . Bring stakeholders to your table. . . .*

What a productive convergence of geo-green activism! In the same news cycle in which the Green Goracle joshed with his climate-crisis message on stage with Leo DiCaprio, the Green Governator spread an aggressive green-is-good-politics gospel among state chiefs gathered in Washington, and in Texas a team of cagey traders—call them the Green Accommodators—acquired a high-polluting utility in a way that evoked the applause of stakeholders who saw both green and gold. Looking for lessons you can use?

How about a benchmark for comebacks and message consistency?

1. Don't give up on the strength of the executive or the message. . . .

Al Gore stayed on message for six years.

It's not easy for a lifetime politician to climb back onto the ladder after a fall, but the enterprising Gore found, as each business enterprise seeks to find, the right way to reposition a product that cannot be justifiably abandoned.

In 2001, Gore was encouraged by green activists—notably Natural Resource Defense Council supporters in the film business—to rejuvenate his 1990s' Earth-in-the-balance message.

Following the political and business principle that you can't win all the people (customers, investors, film producers) all the time, he targeted. With a one-to-one stakeholder strategy, he went where directed, saying the same thing to similar people who were telling each other this is good and worthy and actionable.

A Washington writer described the transformation from "Willy Loman . . . to the Green Avenger"—a downtrodden, lonely traveling salesman growing to a sort of super-hero leading the fight against the doom that he alone fully comprehends.

Down but not out. Not elected but ineluctable.

2. Remember the power of repetition. . . .

Gore took one point—there's a climate crisis that we must address now—and he made it again and again, at venues large and small across the country and around the world, tired but determined.

Gain trust so you can manage to his advantage. . . .
One of his handlers, who saw the potential for a campaign toward commercial theater success in Gore's 100-slides show, said with some amazement that the former Vice President of the United States "did everything" he was advised to do.

3. Explode any incipient criticism. . . . Grace works. . . .

Gore anticipated the possibility of a "loser" label. His customary self-introduction before each show on the road—"Hello, I'm the former next President of the United States"—was more winnable than wince-able.

Likewise, Governor Arnold Schwarzenegger, arriving in Washington in time for a White House reception on Oscar night, said in a speech the following day that he had erred in pushing California initiatives in a failed special election. "The idea was good but the way I went about it was the wrong way," he said. "I don't claim to be Gandhi."

4. Make the most of the boss' travel. . . . Set the agenda. . . .

Staffs of both Schwarzenegger (Republican) and Arizona's Janet Napolitano (Democrat) had set up speeches for their executives at the National Press Club, distinguishing them from the huddle of governors in town for their annual conference.

Chapter Six Changing Orientation:
Sustainable Course for Corporate Communicators

Schwarzenegger's climate-change message was of his state's leadership and setting standards for carbon constraint—greenhouse gas rollback to 1990 levels by 2020, carbon-dioxide emission limits for automobiles, cap-and-trade mechanisms.

Over at the National Governors Association, the leaders of four states copying the Golden Green State's regulatory formulas would hear Schwarzenegger describe the road ahead as a "signal to the federal government" that if Congress and the Bush Administration don't take the lead on global warming, California and its partnering states will.

5. Bring stakeholders to your table. . . .

Texas was at the same time attracting a cast of partners to pull off the green game of the year—the transfer of a $10.4-billion-a-year energy company, whose future success is based on burning coal, into the hands of new owners who bet on both ample clean power and high profit.

Shaping the $32 billion deal on Dallas-based TXU, announced as Oscar statues found shelves in the homes of winners (Gore's producer took his), were mega-stakes buyout artists like Henry Kravis and Stephen Schwarzman, public pension funders and environmental executives like David Hawkins and Bill Reilly.

Hawkins is with the Natural Resources Defense Council (the same activist group that had much to do with Gore's success). Reilly, head of the U. S. Environmental Protection Agency in the Bush 41 administration, is an investment executive who focuses on green business.

6. Figure out their "what's in it for me?" . . . .

Shareowners look for a nice bump in stock sales. The TXU chief executive will reap a sizable payout and stay at the helm after the takeover. Environmentalists have promises of clean-ups and shut-downs and carbon-war research.

Jim Baker, the former Secretary of State, has put his grey-eminence sanction on the deal—and praise abounds from Sen. John Kerry (who nonetheless plans to hold a hearing on it) and Jeremy Rifkin (the veteran corporate critic on genetic modification and president of the Greenhouse Crisis Foundation).

Rifkin seemed to express the benediction, appropriate to the green grouping of wins and win-wins, when he told a *New York Times* reporter, "God bless this takeover of TXU this week."

A week of lessons from events green and glamorous—but withal, just a week. Will the Gore green recognition translate into political power? Will the California green thrust be parried by congressional rationales? Will government regulators (or another private-equity group) break up the green-heart-of-Texas party?

Ask the veteran PR chief in the everyday rocking and rolling, warming and cooling of corporate communications and expect to hear the No. 1 hard-knock lesson:

7. Don't relax. It's not over until it's over. . . .

Looking for 10 Million Climate Activists: The Al Gore 'We' Campaign

The Alliance For Climate Protection, former Vice President Al Gore's green advocacy group, launched its "We" campaign, a $300 million effort to mobilize public opinion on climate change, in 2008. "We are more than a million strong, from across America," said a narrator in the initial TV ads. Six pairs of people sit down together on a couch, holding signs that express their differences. "Burgers" and "tofu," reads one pair, while others highlight "blue collar" and "white collar," "East Coast" and "West Coast." The pairs are physically very different, and the spot closes with a young white boy and an older black woman sharing a sign that simply says: "We." Tagline for the campaign kickoff ad was: "You can't solve the climate crisis alone, but together, we can." A Gore video opened the Web site in 2008, with the words: "America must commit to producing 100 percent of our electricity from renewable energy and other clean sources within 10 years." Here is an excerpt from the Alliance Web site:

We Need to Act Now

The international scientific community agrees that we have only a short time to act in order for the next generation to inherit a healthy planet. And while public awareness of climate change is now high, a sense of urgency and an understanding of the solutions needed remain alarmingly low. Climate change is still largely seen through partisan filters and advocates of action too often must fight entrenched ideology and cultural stereotypes. Breaking the partisan gridlock will require Americans of all political stripes to call for bold action, which will only happen when solving the climate crisis becomes a moral imperative instead of a political issue.

We Can Solve the Climate Crisis

There are many effective players fighting to solve the climate crisis on all levels, and the solutions needed are at our fingertips. We know that we need to act and we know what needs to be done. What has consistently been missing, however, is a massive and sustained national effort to catalyze a broad culture shift on the issue, raising the climate crisis out of a partisan framework and unlocking the potential for real solutions. Our leaders will take the bold actions needed to solve the climate crisis only when the American people demand that change. Through a robust paid media campaign, cutting-edge online activation and partnerships with mainstream civic and religious organizations, the Alliance has set a goal of enlisting an unprecedented 10 million citizens to become climate activists.

Source: Alliance for Climate Protection, http://www.wecansolveit.org

"'License to operate' can no longer be taken for granted by business as challenges such as climate change, HIV/AIDS, water scarcity and poverty have reached a point where civil society is demanding a response from business and government."
—*Al Gore and David Blood*[46]

"We believe it is useful to share with our investors and our customers our plans and progress in addressing climate change and other sustainability issues. They should know where we stand. Our approach is based on our heritage of democratizing technology. Our blueprint for sustainability makes it clear our approach is based on affordable, volume related technology that can affect millions of vehicles."
—*John Viera, Ford Motor Company*[47]

---

[46]*Wall Street Journal, OpinionJournal*, March 28, 2006. Former Vice President Gore co-authored this opinion piece in his role as chairman of Generation Investment Management (GIM). Mr. Blood, formerly head of Goldman Sachs Asset Management, co-founded GIM with Mr. Gore.

[47]Mr. Viera, Director, Sustainable Business Strategy, Ford Motor Company, quoted on SocialFunds.com on April 15, 2008

Chapter Seven

Climate Change Raises Pressure: Investors' New Focus on Sustainability

Early in the new century, a group of institutional investors managing $10 trillion in assets got together and decided that global warming was enough of a shareholder issue that they could put organized pressure on major companies to communicate more openly and consistently about their climate change exposure. The resulting "Carbon Disclosure Project" involved questionnaires sent to 500 of the world's largest companies (including airlines, automobile manufacturers, insurers, power generators, retailers, steelmakers, and technology companies) asking them to explain their greenhouse gas emissions policies and strategies. The project then publicized the response (or lack of one) for investors to note.

Today, virtually every large company has responded, or is getting its information together so it can respond, to this now highly-respected evaluator. Now with the strength of major institutional investors managing close to $60 trillion in combined assets, the CDP can rightfully boast that it is the gold standard for carbon disclosure methodology and process, with a Web site bulging with the largest repository of corporate greenhouse gas emissions data in the world. In 2008, more than 3,000 companies had submitted information on the business risks and opportunities presented by GHG and climate change factors.

If any stakeholder or evaluator wants to know a large corporation's climate change and sustainability strategy, a search of www.cdproject.net is the place to start.

Communicate Sustainability With Investors

While corporate people may wince at the implications of "disclosure"—as in being forced to admit to something that was hidden—there is no denying that transparency is a fact of business life and sustainable business communications, and it starts not necessarily with government but with the financial community.

In Greening 2.0, it has become obvious that in order to engage with the true power brokers in the climate change arena, corporate communications must be directed steadily toward the financial community. Investors have coupled with the interests of an expanding universe of players—social-issue groups, free marketers, environmental and energy businesses, as well as major players in the investment community—to question corporate performance, and commitments and risks related to global warming. For the publicly held company and its communicators, this is now an ongoing area of information interchange with financial stakeholders.

"Just like the tipping point that many scientists point to in the Earth's changing climate, investor interest in how to play global warming seems to be gathering steam," said *Wall Street Journal* reporter Keith Johnson, writing as big investors gathered in New York for a 2008 conference on the business impact of government decisions.[48] More than fifty shareowner resolutions demanding climate-change commitments had been filed in the proxy season. Three big banks had just declared they would toughen financial standards for coal-powered utilities. Others had recently launched global-warming indexes and funds that track carbon-sensitive companies. And many Wall Street power players were active in the then $40 billion market trading carbon-dioxide emissions rights.

At the 2008 meeting held under the United Nations umbrella, the participation was high powered. The California Public Employees' Retirement System (the nation's biggest pension fund with major clout in the financial community), the state treasurers or comptrollers of eleven states, and the president of the AFL-CIO took part, agreeing to invest $10 billion over a two-year span in green technologies such as wind and solar energy. The AFL-CIO leader, John Sweeney, told the gathering that union workers' retirement funds need to be invested in the war on carbon. "These deferred wages of working people are the capital that can fuel the energy economy of the future," he said.

---

[48]"At U.N., Investors Try to Divine How Soon U.S. Will Go Green," by Keith Johnson, *Wall Street Journal*, February 9–10, 2008.

Chapter Seven Climate Change Raises Pressure:
Investors' New Focus on Sustainability

NEW GREEN LANDSCAPE

If you are looking at the sociopolitical scene from the top of your organization, you see Greening 2.0 in the light of climate change factors affecting investor relations. Energy issues have fused with environmental and other social accountabilities to create new financial dynamics. If your company is vulnerable because of legacy issues such as power supply or generation, if you've bought into business areas that bring new carbon accountability, if you're associated with a business group that puts you into play with new stakeholders or offends some of your traditional base, if—in short—you've exposed investors to more risk, or even the perception of it, you can expect investor activism under the flag of carbonomics.

The level of confrontation varies from company to company, depending on the interests of established investors and those drawn to the company with a social activist agenda. Pressure on the company and its board may be relatively modest: request for time to express views at annual meetings, public statements critical of companies either joining or failing to join certain groups or coalitions, blog mentions that put the company in a negative light, references at conferences and panels, and mentions in financial coverage and comments on radio talk and cable TV shows. On the Web, daily coverage of business and climate change is at a consistently high pitch, with sites such as E&ETV and *ClimateWire* becoming must-access for corporate communicators who need to track the fast pace of climate/sustainability developments affecting business.[49]

Ratcheting up the global warming activism can, and will, include shareholder resolutions, demands for board representation, and perhaps the possibility of personal and professional surprises reminiscent of the type we saw in the early years of green activism. I recall the Wilmington, Delaware, billboard with DuPont's chief executive demonized in the chlorofluorocarbon/ozone issue, and I saw a group of shocked executives of another company when activists stood up in the annual meeting and handed a "dirty globe award" to the CEO, taking photos for the newspapers. Will we see these kinds of attacks going forward in carbon war conditions? Well, we did see BP aggressively confronted—and at least its flower-like logo demonized in a drawing accompanying a *New York Times* op-ed written by a

---

[49]*ClimateWire*, billed as the "politics and business of climate change," is edited by John Fialka, former *Wall Street Journal* reporter; accessed at http://www.eenews.net/cw/

disenchanted advertising person—following its energy and environmental lapses. And on any day, it's no problem to find anti-corporate columns, comments, and electronic brick-bats on blogs and Web sites. And while we may not see combative forces climbing over company fences, putting up nasty billboards, or storming an annual meeting to hand the CEO a dirty memento, a visit to YouTube will put us into the real/virtual world where anyone is vulnerable and on view 24/7. An executive caught on a cell-phone camera in an anti-green moment—a speech, an encounter at a meeting or trade show—can be placed into a grim groundhog day reliving for as long as there's Web space.

COMPANIES AND SHAREHOLDER RESOLUTIONS

Citigroup, JPMorgan Chase, Morgan Stanley, and Merrill Lynch have all published research reports analyzing the financial performance of the carbon markets with preliminary evaluation of corporate leaders and laggards in their various sectors. Some encourage investment, such as Merrill Lynch's spotlight on companies best position to capitalize on "the clean-car revolution," while others note vulnerabilities.

Shareholder resolutions, putting pressure on company management to address both the risks and market opportunities associated with global warming, are on the rise. The 2008 proxy season filings of 50-plus climate-change proposals (up from 39 the previous year, and 30 the year before that) followed the escalation of global warming news, state and federal government action and reports—on Web sites—from companies on climate change and sustainability. Critics and admirers are tuned in. According to the nonprofit Social Investment Forum, "concern over climate change—and whether companies are preparing strategies for the geophysical, regulatory, and litigation risks it poses—constitutes a top and closely watched category of shareholder proposals."

Beyond the large nets cast by outfits such as the Carbon Disclosure Project that reach thousands of major companies worldwide asking for information in a generally consistent format, many hundreds of other companies of all size are routinely asked by company investors for specific information of narrower gauges. Companies are called to task—or at least to respond—on sociopolitical and economic implications, seeking proof that management's business decisions have taken into account the relevance of climate change. Investor questions most often focus on energy efficiency measures, greenhouse gas emissions reduction, and the development and use of renewable energy sources.

Three of the shareowner proposals in 2007 were aimed at ExxonMobil, occurring as activist investors rode on the wave of criticism that was rolling when Congress began its session and Senator Barbara Boxer, chairing the Senate environment committee after the Democratic takeover in the 2006 elections, blasted Exxon for prolonged support for scientists skeptical of the cause and effects of global warming. At the 2008 annual meeting, Rockefeller family shareowners confronted ExxonMobil management with climate change resolutions that, while failing adoption, stimulated negative media attention. The company subsequently launched an aggressive advertising campaign on its green energy commitment.

Avoiding Resolutions

Companies routinely try to head off proposed global warming resolutions. SIF observed that "pressure from concerned stakeholders has produced significant victories," and offered ConocoPhillips as its case in point, noting that the company had announced it was joining the U.S. Climate Action Partnership, the group of large companies advocating early government action to reduce global warming. ConocoPhillips' funding of an eight-year, $22.5 million biofuels technology research program at Iowa State University was the sign needed for an SIF member, Trillium Asset Management, and its co-filers to withdraw a proposal demanding company action on renewable energy alternatives.[50]

The Service Employees International Union also withdrew a resolution proposed for the Wells Fargo 2007 annual meeting when Wells Fargo agreed (after the proxy statement had been printed) to do a climate assessment in three key sectors of its lending portfolio and to share the findings with SEIU. The drafted resolution asked the company to formulate "comprehensive greenhouse gas emissions reduction goals for its own operations and those of its corporate borrowers."

Wells Fargo, one of the country's leading purchasers of renewable energy offsets, has staff specialists studying the implication of climate

---

[50]Social Investment Forum news release, "Investors Filing Record Number of Social and Environmental Shareholder Resolutions in 2007: More Disclosures Sought on Climate Change, Political Contributors and Sustainability at Leading U. S. Corporations," May 30, 2007, http://www.socialinvest.org/documents/2007ProxySeasonPreview_001.pdf (accessed 6/07/07).

change on its businesses. The employees union sought the shareholder resolution, a spokesperson said, because "we want them to rethink their business, and set themselves up to take strategic advantage of climate change." The group wanted to know if Wells Fargo was lending money to companies that could be forced into bankruptcy because of greenhouse gas regulations, and if the bank was financing new technologies for alternate energy or offering climate-change consulting services to customers.[51]

General Motors shareholders considered a proposal from the Connecticut State Treasurer's office and more than a dozen members of the Interfaith Center on Corporate Responsibility (275 faith-based institutional investors) calling for GM to set greenhouse gas reduction goals for its operations and vehicles, and to deliver to shareholders the plan agreed on by management to reach the goals.[52] The unsuccessful shareholder proposal was put into an altered perspective by the time of the GM annual meeting; the company had joined the USCAP and was associated with its carbon constraint goals.

Other companies and their recent shareholder challengers include Bed Bath & Beyond, pursued by the Sierra Club Mutual Fund; ACE Insurance, by the Calvert Group; and Dominion Resources, the electric power and natural gas company, by New York City Comptroller's Office. "It is incumbent on us as trustees of pension funds," commented a representative of the New York comptroller, "to find out how companies are mitigating risk to our investments from climate change."[53]

SHAREHOLDER ADVOCACY TREND CONTINUES

Ceres, the advocacy investing group, helped to coordinate the filings of the 54 global warming shareholder resolutions with U.S. companies during the 2008 proxy season—nearly double the number filed two years previous—targeting electric power companies, oil and coal producers, airlines, and homebuilders.[54] Resolutions sought greater disclosure on

---

[51] Article in the *New York Times* on February 13, 2007, by Claudia H. Deutsch, "Companies Pressed to Define Green Policies." http://www.nytimes.com/2007/02/13climate.html (accessed 2/14/07).

[52] "GM Holders to Eye Greenhouse Gas Proposal," June 4, 2007 article, *CNNMoney.com* (accessed 6/4/07).

[53] Claudia Deutsch article, *New York Times* 2/13/07.

[54] *SustainableBusiness.com* News, March 10, 2008.

Chapter Seven Climate Change Raises Pressure:
Investors' New Focus on Sustainability

the companies' responses to climate change, including greenhouse gas reduction and renewable and energy efficiency strategies. The nation's largest public pension funds, as well as labor, foundation, religious, and other institutional investors were in the game seeking company response. The California State Teachers' Retirement System, the nation's second largest public pension fund, filed climate-related resolutions for the first time. This time, Ceres reported, 14 resolutions were withdrawn by investors after the companies agreed to disclose potential impacts from emerging climate regulations and strategies for reducing greenhouse gas emissions.

Ford Motor Company investors concerned with climate change and a carbon-constrained economy presented climate-related shareholder resolutions to Ford in 2008, but the resolutions were withdrawn after Ford announced it would release its GHG reduction plans. Ford is acknowledged by groups such as Ceres as the first U.S. automobile company to publicly release its plans to meet earlier published goals of reducing its greenhouse gas (GHG) emissions by at least 30 percent from its new vehicle fleet of light-duty passenger cars and trucks by 2020.[55]

ROOTS OF CURRENT ACTIVISM: CERES

Ceres is a predominant player in investor action on climate change, continuing its long-term function as a spark igniting corporate environmentalism. It is a prominent indicator of Greening 2.0's contrast with earlier conditions. Originally known as the "Coalition for Environmentally Responsible Economies," Ceres has since the late 1980s been a watchdog of corporate social awareness. The group's "Valdez Principles" were used to prod business management to improve environmental conduct. The 10-point code, pressed on dozens of companies, was linked to the famous case of the Exxon tanker Valdez, whose accidental release of crude oil in Alaska's Prince William Sound in 1989 was blamed for environmental damage.

With its touch of irony, the name of the proposed code of behavior signaled Ceres' confrontational stance and had the effect of putting companies in a reactive, if not defensive, mode. Some companies, mostly smaller companies with established green-market orientation such as

---

[55]*Socialfunds.com* article by Anne Moore Odell, April 15, 2008, "Ford Establishes Greenhouse Gas Reduction Plan," by Anne Moore Odell

The Body Shop and Ben & Jerry's, immediately signed on to the Valdez Principles. They exemplified the enterprises that saw the pledge as a marketing advantage, using it competitively, while the added value of protection from other environmental activists' demands because they were on Ceres' list, validated as greening companies. Other company managements perceived a pragmatic opportunity to make a commitment to a cause with political implications, seeing in Ceres a group allied with both investors and consumers with the potential for extended influence on public perceptions and public policy.

Ceres found a different story when they took their principles to the management of larger companies. Executives who had been on the front lines of green battles that began in the 1960s were not ready to sign on to the Valdez Principles. Although few would argue with the call for "wise use of electricity" and "environmental risk reduction," or even the appointment of "environmental directors," most felt these and some of the other suggested pledges were internal management matters, and many considered them to be matters already under control. Chemical, oil, mining, automotive, and other manufacturers were making considerable progress by 1990. With the reality of federal overview (EPA had been in business since 1970), corporate environmental management was methodically addressing clean-up, process and waste minimization, and regulatory burdens. Companies were working with trade associations, environmental organizations, and government to develop general as well as industry- and company-specific guidelines.

Green Missions

Green mission statements were issued by several companies. And, in anticipation of the 1992 U.N. Earth summit, when the world's spotlight would shine on American problems and progress in environmental issues, corporate executives were getting ready to announce a global commitment—the idea that would become the World Business Council on Sustainable Development. Executives of companies like DuPont, Procter & Gamble, Dow, AT&T, and Royal Dutch Shell were developing an evaluation method that would enable companies, as well as government and NGOs, to set benchmarks and to chart progress in controlling manufacturing process emissions.

Chapter Seven Climate Change Raises Pressure:
Investors' New Focus on Sustainability

Corporate America was, in short, moving toward a green, sustainable system that would satisfy government requirements and align with changing stakeholder expectations. With this internal focus and collaboration of companies devoted to greening, the pressure from external evaluators was deemed as untimely and largely unwelcome. Companies in dialogue with Ceres pointed to their individual commitment, initiatives and efforts to establish volunteer emission goals, guidelines, and measurement.

Corporate communications were not enough to fend off the first strong thrusts of green shareholder proposals, however, and Ceres led the environmental groups in the early rounds of shareholder proposals and annual meeting challenges. At several companies' annual meetings, stockholders asked management to support the Ceres pledge or to explain their resistance. Most of the company votes for the original Ceres proposals were predictably small, but at the mineral development firm of Kerr-McGee in 1990, the total shareholder vote in favor of Ceres reached a significant 16.7 percent.

For any number of companies, responding to direct contact from Ceres or merely anticipating the possibility of shareholder interest in the new green challenges, environmental communication was kicked into high gear. Companies began communicating heavily about their green commitment, with and without reference to the proposed standards in the growing green community, and made sure the subject was covered before and after the annual meeting season.

VALDEZ PRINCIPLES ACCEPTED

In 1993, following more than a year of negotiations with Ceres, the oil and chemical company Sunoco became the first *Fortune 500* company to accept the Valdez (soon-to-be-renamed Ceres) Principles. Sunoco's leadership triggered a new round of discussions leading to endorsements by other large companies, including GM, Ford, American Airlines, Bank of America, Catholic Healthcare West, and Northeast Utilities. Ceres today states that more than 70 companies have endorsed the principles formerly known as Valdez, and Ceres has worked with numerous other companies to adopt environmental policies and issue performance reports. Ceres' "partner companies" submit annual green reports, engage with investors and others put together by Ceres to work on social responsibility, and advance sustainability in business practices and reporting.

Ceres now takes to the media its list of ten companies that are not, in the group's opinion, protecting investors through adequate global-warming strategies. The group's Ceres' corporate hit list was weighted in 2007 toward large energy companies, including ExxonMobil, Allegheny Energy, Consol Energy, ConocoPhillips, and TXU.

Bottom Line: Uphill Challenges

Groups like Ceres demonstrate the busy bridge that now connects social activism with Wall Street, bringing uncommon change to the once relatively small and often lightly regarded social-investment funds. Mindy S. Lubber, Ceres president, has indicated that investor pressure about climate change is different from social investment evaluations of the past, where company economics were not heavily considered, if at all, in demands for environmental cleanup. "This has nothing to do with social investing," Lubber has said in media interviews. "These investors are owners who want the companies to stop being laggards when it comes to minimizing risk and taking advantage of opportunities."[56]

As in the early greening years, some of the companies on the global warming list expressed confusion as to the reason they had been negatively tagged, while others acknowledged they were probably not communicating their commitments, and actions to address risks, as well as they might. Mary S. Wenzel, vice president of environmental affairs for Wells Fargo, said the company had invested $125 million in renewable energy projects during a previous six-month period and had lent $750 million in recent years to developers of green buildings. "We are demonstrating that we are addressing risk" she said, but acknowledged that company communicators "have not issued the kind of public policy statements that shareholders seem to want."

The option of conforming to a coalition's objectives and guidelines may well be considered as a means of extending both awareness of a company's global warming actions and accountability, and its stakeholder support, avoiding some shareholder concerns and proposals. Certainly, as a coalition open to companies willing to conform, Ceres is a worthy consideration. The group has grown steadily over the years to a reported alliance of more than eighty-five organizations, including the AFL/CIO, National

---

[56] *SustainableBusiness.com News*, March 10, 2008.

Chapter Seven Climate Change Raises Pressure:
Investors' New Focus on Sustainability

Wildlife Federation, Friends of the Earth, and the Union of Concerned Scientists, as well the investor links to the Interfaith Center on Corporate Responsibility, Calvert Group, and Trillium Asset Management.

However, whether contemplating activist or ordinary business groups, companies will need to understand fully the value as well as the possible downside of linkage. Joining business coalitions that are outside the usual industry trade associations can mean compromises that can come back to sting some stakeholders.

CEO Questioned

While Caterpillar was not a target for a global warming shareholder proposal in 2007, CEO James Owen received the punch of a different form of investor activism after the company joined the USCAP, the coalition of companies, and NGOs agreeing on the need for a mandatory cap on greenhouse gas emissions. Prior to the company's annual meeting in 2007, Owens received a well-publicized open letter from a conservative business coalition, the National Center for Public Policy Research (NCPPR) and allies including the Competitive Enterprise Institute and the American Conservative Union, charging that the unintended outcome for Caterpillar shareholders, should such a GHG cap occur, would be lower returns on investments in Caterpillar. NCPPR's David Ridenour, who organized the letter to Caterpillar, told reporters that the company's success depends on coal industry success, since the Peoria, Illinois-based firm is a major producer of mining equipment. "If you have caps that result in less coal production, what's that going to do to (Caterpillar's) orders?" He cited a report from the Congressional Budget Office concluding that a cap on carbon emissions by 23 percent would lower stock values by 54 percent for companies in the coal sector.

Among those signing the letter to Caterpillar was Murray Energy Corporation, the privately-held coal-mining company whose CEO testified before Congress and gave media interviews opposing cap-and-trade and other potential limits to coal mining. Murray Energy notified Caterpillar that it would stop purchasing mining equipment from Caterpillar in protest of its support for cap-and-trade. A representative of Project 21, a conservative organization supporting African-American owned businesses, and one of the sixty signers of the letter to Owens expressed dissatisfaction with Caterpillar at the 2007 annual meeting, saying that carbon caps would increase costs.

At the annual meeting, the CEO explained the economic pragmatism in the company's stand. "We believe that some climate change legislation is likely," Owens commented, "Clean air and less carbon dioxide in the air are going to cost money. We just want to get it done in the most economical way." He said the cap-and-trade system offers a degree of flexibility and cost controls that will help businesses remain competitive. Owens also made a pitch for free trade, warning that Caterpillar, along with other American companies, would be harmed by restrictive legislation. "There's a tremendous anti-trade sentiment and anti-global sentiment, which concerns us," he stated. "The biggest risk to global prosperity is turning inward. You can't build a wall to the year 2050, you've got to build a bridge."[57]

Mr. Owens said he was "disappointed that [Murray] is not going to buying our products anymore," but stressed that businesses can no longer simply refuse to go along with emissions reduction strategies.

Ridenour said that one of the signers of the letter against Caterpillar's stand, the Free Enterprise Action Fund, planned to file a resolution opposing participation in the USCAP at future shareholder meetings.[58]

FORD'S WINNING DIALOGUE

Ford, the company that avoided a climate change resolution in 2008, had previously worked out a modeling process with Ceres and other investor groups to determine carbon emission targets and the steps the company planned to take to reach the targets. The model looked at costs, vehicle technologies, baseline fuels, biofuels and consumer issues connected to the climate concerns of stakeholders. Company spokespersons said the model was not put out as "the answer" but as information on a range of possible solutions that could be understood by interested investors. Ford invited the investor groups to examine the model and to have an early look at the company's planned sustainability report.

Ceres and its fellow investor activists noted the appointment by Ford of a group vice president for sustainability, environment and safety engineering, reporting directly to the CEO, as an indication of Ford's seriousness in making sustainability a high priority. And they were clearly pleased that Ford had shown them the plan to achieve CO_2 reduction goals

---

[57] *Crain's*, article by Bob Tita, "Protesters decry Caterpillar's support of CO2 limits," June 13, 2007.

[58] "Free-marketers Urge Caterpillar to Leave Carbon-Cap Coalition," June 12, 2007, Greenwire online news service. http://www.eenews.net/Greenwire/, accessed 6/13/07.

Chapter Seven Climate Change Raises Pressure:
Investors' New Focus on Sustainability

prior to public release of the company's 2008 sustainability report. "Long term investors need to know that there is a plan in place for our company to be profitable in a carbon constrained economy,"[59] said Sister Patricia A. Daly, executive director, Tri-State Coalition for Responsible Investment, and representative for the Sisters of St. Dominic of Caldwell, NJ, the lead filer of the resolution that was withdrawn.

Ask Not What the Climate Can Do for You, But What It Can Do for Your Portfolio

Investors Discuss How to Stay Wealthy Amid Climate Change

Nearly 500 corporate leaders and institutional investors representing $20 trillion in capital met at the United Nations in February 2008 to discuss the risks and opportunities presented by climate change.

The gathering called itself the largest ever meeting of investment types specifically convened to discuss climate change. Attendees mused about how they could continue to make money in a climate-changed future, set a price for carbon that wouldn't hurt them financially, pressure the U.S. Securities and Exchange Commission to endorse disclosing climate-related risks, and prompt the United States to adopt legislation slashing its greenhouse-gas emissions by up to 90 percent from 1990 levels by 2050.

"This action plan reflects the many investment opportunities that exist today to dent global warming pollution, build profits, and benefit the global economy," said Mindy Lubber of investment group Ceres. "Leveraging the vast energy-efficiency opportunities at home and abroad hold especially great promise for investors."

Attendees pledged to invest $10 billion over the next two years on green tech and to pressure companies to divulge their climate risks.

Source: Grist.org item based on Reuters news story, February 14, 2008.
http://gristmill.grist.org/story/2008/2/14/114523/687?source=daily

---

[59] *Socialfunds.com* article

"What does it matter if one fast food company is singled out as 'best in its class' which is the rationale employed by KLD Research & Analytics? ... If you are going the wrong way, it doesn't matter how you get there."
—Paul Hawken[60]

"If there's something strange in your neighborhood ... If there's something weird and it don't look good—Who you gonna call? Ghostbusters!"
—Song in the 1984 movie

"The majority relies on Greenfluencers to sort through the messaging clutter and 'greenwashing' to determine which corporate claims are truly credible. Consequently, Greenfluencers have significant power to positively or negatively impact sales and/or corporate reputation."
—David Zucker, Porter Novelli[61]

---

[60]Hawken, an environmental activist respected for his early advocacy of tying business aims to social change, founded Natural Capital Institute. This quote is from a 2004 article in *Common Ground* magazine, critical of SRI funds for lack of standards and transparency on screening and holdings for shareowner action. He said the term "socially responsible investing" was broad and meaningless, and called for the common standards, definitions, and codes of practices, presumably of the kinds now in place.

[61]A 2008 study of 12,000 U.S. adults conducted by Porter Novelli, a global communications agency, concluded that a small but powerful group of consumers—"Greenfluencers"—drive trends and shape purchasing decisions in the mass market. Zucker, partner and director of CauseWorks, Porter Novelli's corporate responsibility and sustainability practice, described the study at its launch. More information is available at http://public.pnicg.com/cbintra/public.nsf/39d1dcb34e75dac88825682 7000628ea/9fd93c179b8a00c0852574710060557f/$FILE/PNgreenflu_FINAL%20(2).pdf.

Chapter Eight

Where They Go to
Learn About Your Sustainability

The musical question in the Bill Murray ghost-busting movie—*who you gonna call?*—is one that corporate people wish were easy to answer when it comes to questions about the company. Unfortunately, you know it won't always be you that they call. I like what the public relations community's esteemed patron Harold Burson always said to CEOs who were wondering how to start communicating with outsiders: "Every employee is somebody's expert on your company." Whether it's to get the inside skinny or to check on "something weird [that] don't look good," company stakeholders and would-be stakeholders will turn to their own ghostbusters—family, friends, sales people—who are connected with the company and used to work with the company. They will Google you, they'll heed the media they trust and—the continued focus of this chapter—they will take advice from investment advisers.

Elsewhere in this book, we cover the sustainable communication attention that needs to be given to employees, sales people, suppliers, and others in the chain of opinion and influence. Essentially, these are extensions of basics in internal communication. The company's climate change and sustainability mission, when it is decided, must be clearly understood within the company. Strategies must align with existing core values. Mechanisms for employee involvement in the company's Greening 2.0 decisions, rollout, feedback, and maintenance will be a regular part of the information flow and engagement with other stakeholders. And the company "going sustainable" will have to give its strongest evidence of sincerity to those inside the organization, as well as provide them with case information, talking points, and stories they can relate to because they know about or have participated in sustainability implementation projects. Efforts to cut energy, reduce carbon footprints, develop or use new technology—these kinds of internal commitments will engage employees, families, sales and supply chain people as participants, and ideally, believers, in the company's success, and potentially as positive communicators.

When shareowners and those who advise on investments are confronted with doubt as to whether your company is a winner or loser in the war on carbon, how is your company judged? A company creates long-term shareholder value by managing risks and using opportunities related to economic, social, and political developments. These three legs of the tripod supporting corporate sustainability come into balance in the strategies and communications that appeal to investor stakeholders—and those that influence investments in the social action or NGO community.

How Dow Jones Rates Sustainability

So where do investors go to size up your company and what do analysts tell those who call? The rising availability of company information through mechanisms such as the Carbon Disclosure Project—run out of a London office—are a ready source, but they do not yet rate or rank companies in the way that older line systems such as the Dow Jones Sustainability Index (DJSI) do. The DJSI is North America's premier standard, using a systematic corporate evaluation to identify the leading sustainability-driven companies in the various industry sectors. Criteria today include climate change strategies, energy consumption, human resources development, knowledge management, stakeholder relations, and corporate governance. Of the largest 600 North American companies on the Dow Jones global index, the DJSI picks the top 20 percent in terms of sustainability, adjusting for general as well as industry-specific sustainability trends. Recent top holdings have included GE, Citigroup, Microsoft, Bank of America, Procter & Gamble, Pfizer, Johnson & Johnson, Chevron, Cisco, and IBM.

While "environmental" and "social" factors important in a company's economic success were in the same evaluation mix in the earlier Greening 1.0 years, subsequent developments—not least Sarbanes-Oxley regulations and prolonged, repeated news events that have driven ghostly fears into public perception of corporate responsibility—led evaluators like Dow Jones to the view that the terms should be differentiated.

Chapter Eight Where They Go to Learn About Your Sustainability

Environmental developments are the more traditional eco-factors. The DJSI continues to look at the company's pollution control and waste management, its awareness of and reaction to its ecological risks, and its global warming strategies. Other hold-over greening 1.0 elements that factor into Wall Street evaluation are risks/opportunities related to population growth, availability of natural resources needed by the company and bio-diversity—topics that have been on the sociopolitical agenda as far back as the first United Nations conference on sustainable development.

Social developments have added to the evaluations and are now considered on Wall Street as a leading way to define a firm's transparency and accountability. Living conditions, public health, equal rights, fair trade, gaps between rich and poor—these are considered in the DJSI equation evaluating companies and comparing them to competitors in the financial marketplace.

With those two terms expanded and separated, the third consideration that undergirds Wall Street scrutiny are *economic developments* that test company management's grasp of the effects of events such as climate change and carbon constraints. Evaluators look at strategies addressing risks and rewards in technology and innovation, speed and flexibility, product life-cycles, organizational learning, intellectual capital, and the increasing matter of "virtual" living and working, where outsourcing and employee location/working flexibility come into play.

The bottom line for corporate sustainability in the DJSI perspective is *greater value generated for shareholders*. DJSI and the great number of companies that follow DJSI guidance have the basic belief that if a company is able to point to its tripartite accountabilities—social, economic, and political (vulnerability or favor in government relations)—it will get respect.

Back in the C-suites, communications and an ongoing focus on reputation are thereby critical in building and delivering proof of the sustainability value base. It is important for CCOs to understand not only how *Fortune* decides which companies go into which industry categories and what criteria the magazine uses to rank admired companies within these categories, but to do a deeper dive into the sustainability research scene and become clear about how the company will sit in the perception scale shaped by the war on carbon.

SAM's World

Working with the SAM sustainability research group (an arm of SAM Group Holding in Zurich), DJSI weeds out companies perceived as "not sustainable" and uses peer group input—surveying others in the industry sectors—along with rules-based comparisons to choose best-of-class firms. In addition to the influence of SAM's proprietary online survey in picking the best, there is a very strong impact of communications. DJSI depends on information derived from news media and shareholder analysis, company documents and policies available to investors and the public, and company interaction with the evaluators including personal visits.

Media, Company Input

The SAM survey sets a maximum number of points for each of the questions, depending on the level of relevance to sustainability. The points are totaled, weighted, and compared with other companies within the same business sector to arrive at the company's overall sustainability rating. Criteria weighting has been standardized. There are two buckets, each totaling 100 percent. In bucket one, the three criteria categories each gets up to 33 percent. Bucket two gives general information and sector-specific information equally, with a maximum of 50 percent weights. Examples of sector assessment criteria might include, for the automotive sector, the carbon intensity of each maker's cars; for banking, the integration of sustainability criteria into project finance; for pharmaceutical companies, access to markets and products in development countries; and for the food sector, commitment and success in healthful products and lifestyles.

Socially Responsible Investing Surge

Today, about one out of every nine dollars under professional management in the United States is involved in socially responsible investing—some 11 percent of the more than $25 trillion in total assets under management tracked in Nelson Information's Directory of Investment Managers.

In its March 2008 report on socially responsible investing (SRI) in the U.S., the Social Investment Forum (SIF) found it was growing at a faster

Chapter Eight Where They Go to Learn About Your Sustainability

pace than other investment assets under professional management. Spurred by factors such as rising institutional investor interest, growing demand for climate-related renewable energy alternatives, concerns about the Sudan humanitarian crisis, and the emergence of new products, SRI assets increased more than 18 percent from 2005 to 2007. In the same period, all investment assets under management edged up by less than 3 percent. SIF identified $2.71 trillion in total assets under management using one or more of the three core SRI strategies— screening, shareholder advocacy and community investing—up from the $2.29 trillion documented two years earlier.

The March 2008 comment by SIF's board chair made it clear that the sociopolitical influence of climate change has escalated the level of social investing, and has spread the base of active investors. Dr. Cheryl Smith said in presenting the findings: "Increasingly, money managers are incorporating social and environmental factors into their investing practices, acknowledging the demand for social-investing products and services from institutional and individual investors, socially concerned high-net-worth clients, individuals seeking SRI options in their retirement and college-savings plans, and 'mission-driven' institutions, including foundations, endowments, labor unions, and faith-based investors."[62]

KLD Research & Analytics, put together in 1988 in Boston by Peter D. Kinder, a social-responsibility advocate, lawyer, and writer, produces the Domini 400 Social Index, which screens U.S. equity portfolios on social criteria. KLD's stated mission is to remove barriers to socially responsible investing by providing institutional investors with social research, compliance services, benchmarks, performance analytics, and consulting.

Kinder's articles on socially responsible investing and fiduciary duties have appeared in publications in the US, Canada, the UK, and India. Think pieces such as one entitled "Socially Responsible Investing: An Evolving Concept in a Changing World," are available on KLD's Web site.[63]

---

[62]Social Investment Forum, 03/12/2008 *SustainableBusiness.com* http://www.socialinvest.org/resources/research

[63]Peter Kinder co-founded KLD with Steven Lydenberg and Amy Domini, with whom he has co-authored two books, *The Social Investment Almanac* (Henry Holt, 1992) and *Investing for Good* (HarperBusiness, 1993). With Lyndenberg, he also wrote *Mission-Based Investing* (1999), which was being revised in 2008 for a new edition. From 1973 to 1988, Kinder practiced law, first as an assistant attorney general in Ohio, then in Boston as a staff lawyer for a foundation, and finally in private practice. He specialized in administrative law and corporate regulation. For more information on KLD, see http://www.kld.com.

SOCIAL INVESTMENT TRENDS

SCREENED FUNDS: Assets in all types of socially and environmentally screened funds—including mutual funds and exchange-traded funds (ETFs)—rose to $201.8 billion in 260 funds in 2007, a 13% increase over the $179.0 billion in the 201 tracked in 2005.

Eight socially and environmentally screened exchange-traded funds (ETFs) with $2.25 billion in total net assets were available through the end of 2006—the first time SRI-focused ETFs have been a factor in a Social Investment Forum Trends report.

INSTITUTIONAL INVESTORS: At more than $1.9 trillion in assets, socially screened separate accounts managed for institutional investors and high net worth individual clients constituted the bulk of SRI assets tracked in 2007, up 28% from $1.5 trillion in 2005.

SHAREHOLDER RESOLUTIONS: The average level of shareholder support for resolutions on social and environmental issues increased 57% from 9.8% in 2005 to 15.4% in 2007, a record high.

COMMUNITY INVESTING: Assets in community investing institutions rose nearly 32% from $19.6 billion in 2005 to $25.8 billion in 2007.

Source: Social Investment Forum, 03/12/2008 SustainableBusiness.com http://www.socialinvest.org/resources/research

Carbon Disclosure Project[64]

The Carbon Disclosure Project (CDP) is an independent not-for-profit organization aiming to create a lasting relationship between shareholders and corporations regarding the implications for shareholder value and commercial operations presented by climate change. Its goal is to facilitate a dialogue, supported by quality information, from which a rational response to climate change will emerge.

CDP provides a coordinating secretariat for institutional investors with a combined $57 trillion of assets under management. On their behalf it seeks information on the business risks and opportunities presented by climate change and greenhouse gas emissions data from the world's largest companies: 3,000 in 2008. Over 8 years CDP has become the gold standard for carbon disclosure methodology and process. The CDP web site is the largest repository of corporate greenhouse gas emissions data in the world.

CDP leverages its data and process by making its information requests and responses from corporations publicly available, helping catalyze the activities of policymakers, consultants, accountants and marketers.

In 2008, CDP announced the first results of its Supply Chain Leadership Collaboration, designed to encourage companies to measure carbon risks and liabilities in their supply chains. Founding members included Dell, Nestle, Unilever and Cadbury Schweppes.

---

[64]From its Web site, http://www.cdproject.net as of 4/15/08.

"[Vinod] Khosla . . . the California venture capitalist argued that if cheaper alternatives to fossil fuels are developed, simple economics will ensure their adoption throughout the world. He also insisted that the innovation that will create those alternatives will come almost entirely out of America."
—*The Economist*[65]

"For public companies, going green is hot. Not hot like global warming, but hot for scoring public relations points and, sometimes, profits."
—*Earth Day News Story*[66]

"We have a phrase in the telephone business called the Golden Relay. It's the ultimate mechanical device that's inexpensive and will click on forever and will never need any fixing...When you think of the problems of pollution, defoliation, energy—recycling is at least part of the solution. It may not be the Golden Relay, but it's the closest thing we've got."
—*Marilyn Laurie, Former Executive Vice President, AT&T*[67]

---

[65]Article referencing a debate which *The Economist* helped to organize at Columbia University. Mr. Khosla, Sun Microsystems co-founder, defended the proposition, "The United States will solve the climate-change problem"; in the 7/19/2008 special issue on the energy future.

[66]*Atlanta Journal-Constitution* article by Rachel Tobin Ramos on April 22, 2008 that focused on the use of hybrid trucks at UPS, new lines of products at Home Depot and Coca-Cola's moves to recycle containers and helping theWorld Wildlife Foundation conserve freshwater resources.

[67]*From Camelot to Kent State: The Sixties Experience in the Words of Those Who Lived It*, by Joan Morrison and Robert K. Morrison, published by Times Books/Random House, New York, 1987; issued as Oxford University Press paperback, New York, 2001.

Chapter Nine

Carbonomics:
Putting CO_2 into the Corporate Equation

The golden metropolis of Abu Dhabi, the super-rich center of the United Arab Emirates, sitting on 8.5 percent of the world's oil reserves, does not spring to mind when one thinks of cities aggressively engaged in the war on carbon. According to United Nations Development Program data, UAE greenhouse gas emissions are the third highest in the world. Abu Dhabi's carbon footprint makes virtually any big American city look like a carbon pussyfooter.

However, as perhaps the supreme example of metro moves toward climate change adaptation, Abu Dhabi is on track to become the world's first zero-carbon, zero-waste city, entirely powered by renewable energy.. The UAE, bankrolled by Abu Dhabi's crown prince, is spending $15 billion to create the world exemplar of a clean tech community. But the interesting fact of this commitment is that it is strategy not only to fight global warming, but also to turn the oil wealth to long-term economic, educational, and social benefit.

With 420,000 inhabitants worth on average $17 million each, Abu Dhabi has decided to make plans for the future, to look beyond current favorable economic conditions to a day in which carbon alternatives will gain favor. Abu Dhabi's plan is to develop solar, hydrogen, wind and other technologies, with a graduate school for the study of renewable energy, run in cooperation with the Massachusetts Institute of Technology.

The director of special projects at Abu Dhabi Future Energy Company explained, at a renewable energy conference in 2008, that the crown prince's investment would help to transform the UAE from a commodity-based economy to a knowledge economy. "(The plan) will create a lot of high-quality jobs that will have an economic spillover effect," the director stated. In other words, the strategy now in play in the world's richest metropolis is intended to achieve what a similar strategy can do for corporations in the U.S. It will work with the evident economic, social and public policy realities to create a future sustainability.

One aspect of the Abu Dhabi example is common to those in American C-suites forming sustainable strategies. Abu Dhabi leaders are also involved in renewable energy projects elsewhere around the world, including making private equity and venture capital investments in renewable energy companies outside the Persian gulf region. In this country, companies are looking at their opportunities to develop technologies in other countries that can be contributors to their future success.

AMERICAN CARBONOMICS

In 1987, just when the executives of major corporations thought they had their arms around their accountability for managing waste and pollution, a new deal was beginning. American business was called on to solve a global problem involving a greenhouse gas. While not fully appreciated at the time, this was a threshold event leading eventually to all-out war on carbon with its game-changing effects on markets, products, technology, innovation and corporate cost equations.

It was a remarkably successful set of steps. In the 1980s, scientists had convinced national leaders that chlorofluorocarbons, a class of chemicals commonly used in air conditioning and spray cans, were reaching the stratosphere and destroying layers of ozone that protect the planet from global warming. International negotiations led to a decision in 1987 in which U.S. officials agreed with those in 24 other nations to outlaw CFCs. Major American manufacturers such as DuPont got with the program and worked out a plan to stop making the product despite its proven benefits and essential value to customers. Within a decade, CFC production was at zero and, thanks primarily to American business, the problem posed by this greenhouse gas was solved.

With the signing in Montreal of the CFC constraint protocol, "going green" commitments among American companies no longer were confined to the effects of products and production on the Earth's rivers, land, ambient air, fish and wildlife—as had been the case since America's environmental age began with the big boom of Rachael Carson's *Silent Spring* in 1962. The global attention to a greenhouse gas signaled the run up to today's corporate positioning on global warming, with previous greening models subject to rethinking and tilting in the direction of

Chapter Nine Carbonomics:
Adjusting the Corporate Strategy

carbonomics, when prevailing market economics would be infused with a new level of environmental management that puts heavy emphasis on fossil-fuel energy reduction.

Other moves would follow and reinforce the carbon constraint outcome. By the 1990s, the sociopolitical issue of acid rain—that is to say the windborne fall-out of sulfur dioxide emissions from power plants and other sources—led to a market mechanism for American business and government to deal with emission reduction. Implementing a congressional mandate, the U. S. Environmental Protection Agency's 1991 settlement strategy told companies that in order to cut the atmospheric impact of SO_2 emissions, they could make their individual constraint efforts part of a scheme that would result in an cooperate, somewhat flexible approach. Companies—as well as financial entities—could begin to buy, sell or bank the rights to emit sulfur dioxide. A pollutant would be treated as a commodity traded in much the same way as any other commodity in American financial markets. The first substantial test of the market-based approach known as "cap and trade" was under way.

CCO ROLE IN CORPORATE CARBONOMICS

The chief communications officer is in position to contribute to corporate sustainability strategies that involve reputation, stakeholder, and sociopolitical support.

Listen: Ask questions inside the company, and research external stakeholders to understand what's known and not known about the company's carbon war risks and opportunities.

Learn: Engage with CEO, CFO, and business units to understand the company's carbonomics—cost and revenue factors related to climate change.

Lead: Orient the C-suite relevance of carbonomics to business mission execution through internal and external channel competence, stakeholder-relevant input, carbon market awareness, and two-way one-voice messaging.

At each level, the CCO acts as a steward of stakeholder trust; focused on creating a consistent record as a credible, caring, successful force to address climate change matters.

In 1992, Eastern cities worried about the effects of nitrogen oxides emissions from both mobile and stationary sources, as they tried to bring ozone levels down to federal requirements, began a similar market experiment in NO_2 emissions trading. That same year, world leaders at the United Nation's Earth summit put climate change into play as a proposed global treaty that would become known, for the city in which the protocol was subsequently approved, as Kyoto.

For American based companies, "going green" had begun its shift from a deliberate, 40-year process of managing green concerns about generally accepted pollutants. The characteristics of Corporate Greening 2.0 were a skyward focus, greater shared responsibility as well as new areas of competition with other companies, engaging with suppliers and customers to assemble more effective chains of green control, and—most important—a commodity-like market approach to control. These base points referenced by government and the business community would ultimately work *carbonomics* into decision making and introduce carbon trading.[68] A substantial economic prospect was thus established; the World Bank estimates that emission-reducing projects to be financed by trades will total $12.5 billion by 2012.

LEARNING FROM EUROPEAN EXPERIENCE

Government requirements to make carbon emissions a traded commodity in the U.S. can make the economic leg of the corporate sustainability equation more than a little shaky. Experience in Europe has revealed the volatility that companies can face.

Carbonomics made its debut as a virtual yo-yo in the European Union. With government mandating cap-and-trade rules, a ton of CO_2 traded for as much as $15 and as little as 15 cents over an initial two-year span. Disruptions and inequities occurred. Some company emitters paid dearly, others got windfall income, power customers were socked with pass-through costs, and—ironically, the aim of all this, fighting global warming, was not realized. The government conceded that after its trading scheme's first testing period, the net effect on reducing climate-change emissions was minimal.

---

[68]More recently, the northeastern states have begun implementing the 2005 Regional Greenhouse Gas Initiative to reduce CO_2 emissions from power plants, and California has begun regulations on greenhouse gas emissions required by the state's Assembly Bill 32, enacted in 2006.

Chapter Nine Carbonomics:
Adjusting the Corporate Strategy

The good news for U.S.-based companies is what they can learn from the European experience. As William A. Pizer, senior fellow at Resources for the Future, commented during congressional debate on this in 2007, "You don't want to go broke implementing climate policies," urging Congress to "recognize the economic reality of not needing to jerk everything around too quickly."

Based on the EU experience, Washington legislators on both sides of the political aisle proposed to include in any cap-and-trade mechanism a limit on the top price of traded CO_2.

Utilities, major power users and others in the business community are constantly engaged in favor of government moves to create some certainty about carbon prices to plan projects and to provide some near-term protection to the economy while those projects are being built. In addition to a price safety valve in regulations, Congress has put other social and business interests on the table. These include the idea of sending the revenue from emission allowance auctions to a fund that would give companies incentives for new carbon-limiting technologies and successful plans to capture and store released carbon.

Chicago Exchange's Voluntary Mechanism

While American lawmakers worked on a mandatory approach to carbon trading, a voluntary exercise was warming up carbon market mechanisms among companies headquartered in North America. In 2003, two years before trading began in the European Union's government-controlled system, the Chicago Climate Exchange (CCX) began providing U.S.-based entities—corporations as well as cities and others—a cooperative carbon reduction and trading system. By 2008, CCX had enrolled large companies such as Ford, Motorola, Baxter and Bayer; utilities such as TECO and Green Mountain Power; universities such as Tufts and University of Minnesota; non-governmental organizations such as World Resources Institute and Rocky Mountain Institute; farmers in Iowa and Nebraska; cities such as Chicago, Portland (OR) and Oakland (CA); and at least one state, New Mexico.

Participants in the Chicago exchange sign on to a legally binding multi-sectoral, rule-based and integrated greenhouse gas emission registry, reduction and trading system. Trading options include not only CO_2 but all six of the identified greenhouse gases and employs independent verification of collected data.

In Phase I of its operations, CCX members made commitments to reduce GHG emissions 1 percent per year for each of years 2003–2006, below the average of a baseline period, established as 1998–2001. Reductions are calculated in absolute tons. For government entities, calculations include emissions from city-owned operations only. Indirect emissions are included on an optional basis. Phase II parameters extend the reduction period through 2010, with an additional 2 percent reduction commitment for current members and a total 6 percent reduction commitment by 2010 for new members below an established baseline.[69]

The voluntary CCX effort became a useful tool for companies to understand the price of carbon and the economic effects in their operations, and it moved forward as a probable model for federal mechanisms.

A CCX subsidiary in Europe—the European Climate Exchange (ECX)—offers a platform for carbon emissions trading in the European Union emissions trading scheme (EU ETS), with standardized contracts and clearing guarantees. In 2007, more than 50 companies and banking institutions were engaged in ECX, including ABN AMRO, Barclays, BP, Calyon, Fortis, ICAP, Morgan Stanley and Shell. ECX trading represented 75 percent of the total EU ETS volume.

In January 2008 European Union authorities began adjusting the pan-Europe emissions trading system, putting emphasis on increasing the number and level of allowances that are auctioned off, as opposed to given away—which was the cause of market disruption in the first phase of the EU ETS. Other changes were toward settling on one cap on all EU emissions instead of 27 different caps; as well as redistributing auction allowances from richer countries to poorer countries, to encourage the development of climate friendly technologies in all the EU partners.

REGIONAL CARBON TRADING

Setting the pace for U.S. regional carbon compliance were the first two trades of the U.S. Regional Greenhouse Gas Initiative allowances, completed in the spring of 2008. Both trades were in the range of $5 - $10 per ton, considerably higher than the RGGI officials' original guesstimate of $2.32 per ton. A bullish view of regional trading based on this initial

---

[69]See http://www.chicagoclimateexchange.com for a roster of CCX Members, news archives and current prices. Also, see http://www.europeanclimateexchange.com.

Chapter Nine Carbonomics:
Adjusting the Corporate Strategy

exercise suggests a $1 billion market. However, in regional markets as in national, both in the U.S. and in Europe, the bulls must be balanced with the bearish reality: Forecasts and over-allocation were common features in first runs at cap and trade. Caps were set well above actual emission cuts. RGGI indicated there would be no re-assessment of the cap before 2012, so a high supply of allowances would continue to mean little constraint on the market. The bottom line is that eventually, these regional markets will merge into a federal program, at which point the experience of markets like RGGI should be a useful guide in caps, trades and allowance pricing. Meantime, as was the case with companies hooked up with CCX and with the Europeans, the RGGI companies—who got involved even before RGGI had issued allowances, set up a registry or even agreed on a standard contract—were learning how to get footholds, gain exposure and in fact help make the markets.

Carbon Offsets Aim at Neutrality

Carbon emission offsets—practices aimed at achieving carbon neutrality—bring both flexibility and complications to a company's carbonomics equation. The concept is simple in theory: Carbon that is captured—contained, buried, sequestered—is assumed to counteract the carbon that is released into the air. Since the aim is to reduce the greenhouse gas impact globally, CO_2 emissions from one location can be neutralized by reducing or capturing and holding CO_2 in another location.

Entrepreneurs got into the offsets game shortly after the turn of the century, offering to help consumers and corporations to be good citizens and neutralize their carbon footprints. Tree planting was a popular device. Companies, cities, groups and climate-concerned individuals paid an agency or group to plant trees to offset some level of carbon emissions.

Tree planting has its detractors—including biological and climate scientists—who question the effectiveness of this form of carbon capture. While it is true that trees and other green growing things are highly efficient at absorbing and storing energy in the form of ambient carbon dioxide, it is also true that when leaves fall and plants die back, carbon is again released. Skeptics also doubt the trade-off benefit when trees are planted in certain regions of the world, whether lands are cleared to plant trees, whether the trees are cared for and sustainable, and in fact whether in some cases the trees "bought" through third parties are actually planted.

Major companies understand this and are taking steps to guard against misrepresentations of offset schemes and to focus on obvious, measurable need and results.

Marriott International in 2008 entered into a careful structure with the large Brazilian state of Amazon.comas to protect 1.4 million acres of endangered rainforest as one of the first partnerships between government and the private sector to reduce the very significant threat of greenhouse gas increase linked to deforestation.

IT'S NOT EASY (CLAIMING) TO BE CARBON NEUTRAL[70]

As this article in the *Los Angeles Times* underscores, making a claim of carbon neutrality—and buying offsets—is tricky, and (a lesson for all communicators) cynics with their calculators are ready to expose flaws in the claim.

> In 2007, the Oscar-winning film An Inconvenient Truth touted itself as the world's first carbon-neutral documentary. The producers said that every ounce of carbon emitted during production—from jet travel, electricity for filming and gasoline for cars and trucks—was counterbalanced by reducing emissions somewhere else in the world. It only made sense that a film about the perils of global warming wouldn't contribute to the problem.
>
> Co-producer Lesley Chilcott used an online calculator to estimate that shooting the film used 41.4 tons of carbon dioxide and paid a middleman, a company called Native Energy, $12 a ton, or $496.80, to broker a deal to cut greenhouse gases elsewhere. The film's distributors later made a similar payment to neutralize carbon dioxide from the marketing of the movie.
>
> It was a ridiculously good deal with one problem: So far, it has not led to any additional emissions reductions. Beneath the feel-good simplicity of buying your way to carbon neutrality is a growing concern that the idea is more hype than solution.

---

[70] *Los Angeles Times* article 9/3/07.

Chapter Nine Carbonomics:
Adjusting the Corporate Strategy

> *According to Native Energy, money from An Inconvenient Truth, along with payments from others trying to neutralize their emissions, went to the developers of a methane collector on a Pennsylvanian farm and three wind turbines in an Alaskan village*
>
> *As it turned out, both projects had already been designed and financed, and the contributions from Native Energy covered only a minor fraction of their costs. "If you really believe you're carbon neutral, you're kidding yourself," said Gregg Marland, a fossil-fuel pollution expert at Oak Ridge National Laboratory in Tennessee who has been watching the evolution of the new carbon markets. "You can't get out of it that easily."*

Renewable Energy Offsets

While overuse of the offset notion has led to lack of respect if not ridicule, offsets from alternative or renewable energy sources are encouraged by governments worldwide.

Congressional and state government incentives now involve tax write-offs and grants to raise the use of solar, wind, hydroelectric power and biofuels. Energy conservation techniques such as energy efficient buildings; the use of engines that produce less carbon dioxide; efforts on farms, in landfills, in coal mining and in other segments of the American economy to capture and destroy greenhouse gases are all developing and can come into consideration as renewable offset credits in a company's carbonomics calculations.

When renewable energy credits are bought and traded like a commodity, a typical credit unit represents one megawatt-hour of energy produced by a qualified renewable source, such as solar or wind power. The idea is that each renewable energy credit is assumed to replace or neutralize one hour of "dirty" energy. A company works within an emission trading system, uses the market transaction to reduce its reportable net carbon emissions, and factors this into company financials.

Many companies on the obvious front lines of the carbon war—those producing, burning or otherwise highly dependent on fossil fuels—began banking credits for future deals as Congress got serious about climate change. American Electric Power, the nation's largest coal-fired power generator, agreed to purchase 4.6 million carbon offsets from 2010 to 2017 from Environmental Credit Corporation. AEP's chief executive told its investors that the company was confident that natural global-warming controls were coming, and was building its bank of credit. For early adopters, the offset credit price in the pre-required marketplace was favorable. AEP's deal in 2008 was done at a price estimated at about one fourth the price expected to come under a federal program. Blue Source, which claims to be the leading climate change offset portfolio in North America with more publicly registered third-party verified greenhouse gas offsets than any other company, set up several operations to capture CO_2 from fertilizer plants and sell it to oil fields to boost production. Company executives bought offsets from Blue Source for a premium if they were tied to post-2010 projects, the timeframe in which many companies feel that federal caps will be in place.

Trying to get ahead of the curve, nearly 400 start-ups were operating some 600 carbon-mitigation projects in the U.S. at the start of 2008 and consulting firms were counting on this level to triple within two years. Their primary product: carbon credits. In the financial community, JPMorgan Chase was early engaged along with other investment banks and hedge funds in trading large volumes of carbon offsets, betting that values would soar as federal requirements kicked in. Morgan's offsets business was estimated in the tens of millions in 2007 and projected toward hundreds of millions in years following.

CAPS, TRADE, OFFSETS—PROS AND CONS

The need for carbon calculations to provide business planning was the major reason large companies got involved in Washington. They figured that companies will have to bear most of the first costs of the carbon war and that uncertainty and delay in determining the price of carbon emissions were not in anybody's best interest. There is something like a four-year gap between legislation and regulation. It takes time for Environmental Protection Agency rule-making and to give regulated industry formal notice of required changes. And while government gears up to regulate, to set up its accounting, collection, compliance bureaucracy, affected companies will be able to tell its stakeholders: *we're ready—and we intend to continue to compete on your behalf.*

This is not to say that companies favoring cap-and-trade will escape criticism. Forward moving companies in the U.S. Climate Action Partnership received some heat from stakeholders, including investor advocates who wondered whether asking for regulations was in their best interest.

Chapter Nine Carbonomics:
Adjusting the Corporate Strategy

Holman W. Jenkins, Jr. was sharp in his "Business World" column in the *Wall Street Journal* in January of 2007, saying that companies such as Caterpillar, GE and Duke Energy that endorsed cap-and-trade "… would turn their established habit of using the atmosphere as a free waste disposal into a property right, worth billions. Talk about a low-hanging fruit. They are accustomed to treating carbon dumping as a gimme. Now they'd … get paid for dumping less…"

Advocates of carbon taxes argue that because cap-and-trade relies on market participants to determine a fair price for carbon allowances on an ongoing basis, it could devolve into a self-perpetuating province of lawyers, economists, lobbyists and other market participants bent on maximizing their profits on each cap-and-trade transaction. Opponents of cap-and-trade say the costs, both in government implementation and incurred by companies as more expensive technologies replace older and less expensive coal-fired combustion, would likely be imposed on consumers with little possibility of rebating or tax-shifting.[71]

IS THERE A CARBON TAX IN YOUR FUTURE?

The rise of climate change as an American sociopolitical issue gave a lift to economists favoring progressive tax-shifting to reduce regressive payroll and sales taxes. These economists along with some members of the business community supported calls for carbon taxes to wage war on the production or energy generation in which carbon (fossil fuel) is burned.[72]

Proponents of a flat, carbon tax argue that not only would it be easier to manage and more effective in reducing emissions than a cap-and-trade system, but that it could also be less upsetting to the economy and personal pockets—revenue neutral, if government can assure that collected tax revenue is returned equally to taxpayers. Highest producers of carbon dioxide, such as corporations and higher-income individuals, would be

---

[71] For this view, see The Carbon Tax Center, launched in January 2007 by Charles Komanoff and Daniel Rosenblum, to "educate and inform policy makers, opinion leaders and the public, including grassroots organizations, about the benefits and critical need for significant, rising and equitable taxes on the carbon content of fossil fuels." Komanoff is an energy-policy analyst, transport economist, and environmental activist in New York City. He "re-founded" NYC's bike-advocacy group "Transportation Alternatives" in the 1980s, co-founded the pedestrian-rights group "Right Of Way" in the 1990s, and wrote or edited reports on Subsidies for Traffic, The Bicycle Blueprint, and Killed By Automobile. http://www.carbontax.org.

[72] Under most carbon tax proposals, the manufacture or production of items or material not intended to be used as fuel would not be taxed. For example, products made from plastics, although they derive from carbon, would not be taxed.

taxed the most; and through revenue redistribution, poorer segments of society would get the greater benefit.

The argument has been persuasive in Canada, at least at the province level. British Columbia imposed a tax aimed at cutting three million tons of GHG emissions in five years, and to bring in $1.8 billion in three years by raising prices on all fossil fuels in the province. To enhance its appeal to consumers, the tax was paired with tax cuts and rebates. Critics complained that industrial emissions from oil, gas and cement production were not taxed under the plan and that carbon prices were set at relatively low levels, starting at $10 a ton and going up to $30 by 2012. In Quebec, a carbon tax proposal covered all fossil fuels. Proponents predicted it could raise $200 million a year to fund the province's GHG emissions reduction plans. The ink was not dry on the carbon tax announced by

How Companies Balance Fiscal and Green Sustainability

More companies are looking for ways to be environmentally friendly as the green business trend gains momentum. Finding the right balance between fiscal responsibility and environmental sustainability can be complex for many executives; but sustainability also translates into greater efficiency, which is good for shareholders and activists alike. For example, IBM reports saving over $100 million since 1998 thanks to its renewable energy initiatives, which reduced the firm's carbon dioxide emissions by over 1.28 million tons. Companies looking for similar success should start by conducting their own analysis of various green solutions—some of which may have limited results. Companies can utilize advice from environmental watchdogs or other recognized authorities that have access to superior information and technology. Additionally, experts note that an association with a reputable environmental organization will improve a firm's image. For instance, Chevron teamed up with the U.S. Department of Energy, MIT, and UCLA; while Citigroup and FedEx are among the dozen *Fortune* 500 companies working with Environmental Defense. In the long run, experts predict the United States will regulate carbon dioxide emissions, meaning organizations that start reducing emissions now will hold a considerable competitive advantage in the near future.

—*From* Chief Executive Magazine, *February 2008,* http://www.chiefexecutive.net

the BC provincial government before harsh reactions began. A trucking company CEO predicted that the taxes would suck the lifeblood of local economies by driving up costs in food, merchandise, parts and equipment delivered within the province.

NIGHTMARE VISION IN WASHINGTON

There is a problem, however, when you put this into U. S. sociopolitical contexts, with emphasis on the political. The mention of a new carbon tax would mobilize companies in every part of the country and a lot of end-users of power and fossil fuels. Proposals for carbon taxes bounce around Capitol Hill periodically, but generally succumb to the prospect of organized opposition and voter enmity stirred relatively easily in most congressional districts. President Clinton failed in his BTU attempt (to tax fuels based on their heat—or British thermal unit—content) when oil companies, manufacturers and farmers combined to attack the effort as regressive. As Congressman John Dingell, chairman of the House Energy and Commerce Committee, said to a gathering of business people in Detroit as he prepared to lead the 2007 movement toward an achievable cap-and-trade legislative goal and warning that going in the wrong direction would only shut down the process: "Many members of Congress remember only too clearly the letters B, T and U."

It is a delicate balance for elected officials in the U.S. to legislate carbon controls and revenue raising, and the difference between a tax and a market-based system such as emissions trading is much less when the market contains a price ceiling as a form of "safety valve" that protects traders from extremes. As noted by researchers at Resources for the Future, who helped produce a congressional budget office report in 2008, carbon taxes may well be the route favored by economists as rational and sound, but in the pragmatic light of politics they appear to be destined as an idealistic curiosity. That was the case as the ranks closed behind a cap-and-trade solution that would involve federal, regional and state implementations. By the spring of 2008, at least 30 states were engaged either as participants or interested observers in some form of cap-and-trade program.[73]

---

[73]Among other sources, Debra Kahn's article on *ClimateWire* (3/12/08) was helpful in interpreting the CBO report. ClimateWire is part of EENews at http://www.eenews.net/climatewire.

When Congress took up climate change and energy legislation in 2007, Dingell shook up his colleagues with a nightmare vision of the probable impact (and thereby, without so saying, political fallout) of a carbon tax bill. He announced that he was drafting such a bill. He posted a summary on his official Web site to solicit public feedback. This shock to the legislative system was loaded with specifics of how a carbon tax would hit American pocket books—not dwelling so much on the impact on business—but on individuals. Which of course is to say to Capitol Hill politicians: on voters. The Dingell plan would impose a $50-per-ton tax on carbon and a tax of 50 cents per gallon on gasoline and jet fuel. And it would remove tax deductions for larger homes because of their larger contributions to GHG emission. Mr. Dingell, Democrat of Michigan, asked for public comment.[74] He received an abundance.

The popular environmental Web commentator, *Grist.org*, viewed the move this way:

"A carbon tax is beloved by economists and other wonks as the most transparent, efficient means of cutting greenhouse-gas emissions. Voters, however, tend to hate the idea, and thus most politicians do as well… Dingell, who has served in Congress for 52 years and chairs the powerful House Committee on Energy and Commerce, has been accused of pushing 'political poison' in order to torpedo other climate bills that include boosts in CAFE standards. He denies it, but then again, he says this: 'I'm trying to have everybody understand that this is going to cost and that it's going to have a measure of pain that you're not going to like.' The man sure knows how to excite voters!"

In the same timeframe of Dingell's positioning of the policy alternatives, a major figure in the oil business was expressing a view with a similar touch of irony. In a speech before a London audience, Rex W. Tillerson, chairman and CEO, ExxonMobil Corporation, said "… An upstream cap-and-trade system—that is, a system placing a limit on carbon at the point where the fuel enters the commercial world rather than at the point of emission—offers potential advantages in terms of efficiency and simplicity. It reduces the number of regulated entities and provides a cost of carbon to the entire economy. Similarly, a carbon tax could enable the cost of carbon to be spread across the economy as a whole in a uniform and predictable way. Of course, all these policy options carry significant challenges as well as potential benefits, and the devil is very much in the details."

---

[74]John Dingell's carbon tax message, adapted from his Web site, appears in the Appendix.

Chapter Nine Carbonomics:
Adjusting the Corporate Strategy

Dingell had exposed the devilish horns of the carbon tax alternative and, while cap-and-trade was not yet on the agenda for congressional voting, the House committee chairman was subsequently able to work with other members of Congress and the auto companies in approving the most stringent upward requirements of CAFE—that is the corporate average fuel economy regulations that encourage smaller, lighter, more fuel efficient vehicles—in more than 30 years. It was this consensus move that constituted the major impact of the nation's first significant climate change—dubbed "energy"—bill, enacted by Congress in 2007.

Chief Communicator's Focus

Congress will consider proposals beyond the current energy-impact federal excise taxes that motorists already pay to put fossil fuel in their cars and trucks. However, because of public resistance to taxes and the head start that cap-and-trade has achieved, congressional approval of an across-the-board carbon tax now or in the foreseeable future is highly unlikely.

The C-suite communicator's focus in climate change can therefore be reasonably based on the assumption that caps and market trading of emissions are in the cards, and that each company will need to factor in this new reality. In a larger, more general sense, the communications orientation will be no different from what it is in any other significant sociopolitical matter that stakeholders care about: to assure that the company demonstrates a credible, caring, intelligent plan to be part of the answer as well as a winning competitor in financial performance. The chief communications officer has three levels of engagement in the company's war on carbon: understanding the stakeholders' mindsets, shaping executive understanding of stakeholder impact and suggesting strategies, and executing effective two-way, internal and external information flow.

As the steward of stakeholder trust that comes from understanding the company's plans for success, the CCO needs to engage with others in corporate management and the business and research units to understand the exact nature of the company's carbonomics—the cost and revenue factors related to climate change. The company may not yet have sorted out all the questions of power sources, operational energy efficiencies, product impact and opportunities, but the point for the CCO is to be in on the executive exploration, to learn about this part of the necessary interaction with stakeholders.

The second level is to contribute to the sustainability strategy with input on reputation, stakeholder and sociopolitical support. This can involve CCO research into how other companies are positioning and communicating on climate change and corporate sustainability strategies. Benchmarking other companies, becoming familiar with what's happening in Washington, in the states and cities where the company has major business interests, specifics about investor interests—the entire range of factors that can influence sound decisions in establishing the company's plans will need to be on the table to orient C-suite carbonomics discussions to business mission execution.[75]

The third level is execution. My only reminder here is to stay part of the carbon war action, helping management to do its number one job, which is to create stakeholders in the firm's success. It's now a matter of putting this into the new contexts. Understand who your stakeholders are and who they could be if—as is highly likely in many companies—carbonomics brings on changes in your customer and investor base. Stay engaged in terms that they will understand.

And, while it's probably pretty obvious since we are talking about sustainability, the CCO can advocate maintenance and continuous improvement. This is not a flash in the pan sociopolitical issue. It's building a business that will last, with social and political acceptance, under uncommon changed conditions. The company will need to create and use systems for internal and external feedback in the corporate sustainability program, to keep it fresh, on track and relevant to the business and to each group of stakeholders.

---

[75] A possible step toward benchmarking is provided in the *Guide to Corporate Climate Change and Sustainability Positions,* beginning on page 177.

Chapter Nine Carbonomics:
Adjusting the Corporate Strategy

How the EU Cap-and-Trade Scheme Works

European Union officials set an overall cap or maximum amount of emissions per compliance period. They then give each company in the scheme a CO_2 emissions allowance—which becomes the company's emissions target (or cap) for a compliance period. Each government allowance permits the holder to emit one ton of CO_2. At the end of the period, company emission levels are reported. Each company must reconcile its allowances with its emissions during the period. If emission levels are below the established cap, the company has allowances to sell. If emissions are higher than the cap, the company must purchase allowances from companies which have come in under their emissions reduction targets. If an operator does not hold sufficient allowances to meet its total emissions cap at the compliance date, a financial penalty (initially 40 euros, rising to 100 euros in the second phase of the scheme) is charged for each excess ton.

"Sustainability reporting has come a long way since its origins in the social reporting requirements enacted by the Netherlands and France in the 1960s and '70s. The practice has continually evolved—becoming more robust, more demanding, and more credible as a means for a company to convey the full range of its social, environmental, and economic commitments."
—***Jim Sloan***[76]

"The good news is that more and more companies are releasing annual sustainability reports; the bad news is that this makes differentiating yourself from the competition more difficult."
—***PRNews***[77]

"The Carbon Disclosure Project [is] ... a quiet, mature, expert way to raise the consciousness of CEOs who are under pressure to provide immediate profits for the next quarter's reporting ... [it] is a way to give them the sense that looking at the long term is really the smart and sensible way to go."
—***Trustee of the New York State Common Retirement Fund***

---

[76] Jim Sloan, counselor who has helped many senior executives on sustainability and other corporate matters, provided many helpful insights for this book. Jim's comprehensive article on sustainability reporting may be accessed through http://www.envirocomm.com.

[77] "Sustainability Reporting: Green Is the New Black—Pass It On," *PR News*, October 1, 2007.

Chapter Ten

Communicate Your Carbon War Strategies: Companies Post their Climate Change Positions

When General Electric expanded the bounds of its transparency by issuing its first-ever "Citizenship Report" in 2005, a *Wall Street Journal* commentator called it an appeasement to corporate critics, "subsuming the primary mission of the corporation—making profits—within a host of other goals, ranging from the promotion of biodiversity to the protection of indigenous rights."[78]

In fact, GE was executing a strategic decision to join the mainstream of sustainability reporting as a means of controlling the quality and availability of social-impact information.

The source of the concept of green or sustainability information flowing from companies to publics goes back to the early 1990s when a group of us, led by IBM, developed a voluntary mechanism that we branded as the Public Environmental Reporting Initiative. PERI was a set of guidelines for a company to identify components for comprehensive reporting on environmental performance. These included an organizational profile, a summary of the company environmental policy, data on environmental releases, and a review of product-stewardship efforts. Suggestions were offered for each component. The guidelines were designed to be adaptable to a wide variety of industries, with each industry or company able to choose applicable components. Hosted by IBM's Web site, this early resource has given way to a great many voluntary, as well as commercial, devices to help companies do one thing: Take charge of their information flow, to make it sustainable, understanding and useful to its stakeholders and its critics.[79]

---

[78] Alan Murray, "Will 'Social Responsibility' Harm Business?" *The Wall Street Journal*, May 18, 2005.

[79] Dr. Terry F. Yosie, now president and CEO of the World Environment Center, led my consultancy's team that initiated PERI, working with IBM and other company representatives, soon after the 1992 UN Conference.

Greening 1.0's voluntary initiatives of emissions disclosure have become the more collaborative sustainability reporting mechanisms of Greening 2.0. Building on the historic first reporting begun when manufacturers at the 1992 United Nations conference in Rio de Janeiro announced its voluntary plans to disclose emissions information, companies under the gun in the war on carbon have been particularly aggressive in telling their climate change and sustainability stories.

By 2005, the use of sustainability reporting had skyrocketed among the largest multinationals, with 68 percent of the Fortune Global 250 companies issuing a report.[80] By 2007, the practice had expanded to 67 percent of the Global FT500. And a study issued in November 2007 found that while just 44 percent of the DJ STOXX Global 1,800 index issued sustainability reports, four out of five of the largest 10 percent of companies had done so.[81]

In 2005, KPMG noted that among the top 100 corporations in 19 countries, approximately half of those in the utility, oil and gas, chemicals, mining and forestry-products sectors produced corporate responsibility reports, compared with only 22 percent of companies in other sectors, such as trade and retail.[82]

WHAT CONSUMERS EXPECT FROM COMPANIES

Consumers distrust all industries to a significant extent, but some fare better than others, according to a McKinsey global survey of consumer attitudes towards business. Consumers expect large companies in specific sectors to address sociopolitical problems such as climate change and the affordability of drugs in developing countries. Moreover, the survey reveals large variations in what consumers in different countries expect from different industries. The findings suggest that to win the public's trust, companies should tailor their approaches to suit the realities in their own industries and in the countries where they operate.

Source: McKinsey, August 2007

---

[80] KPMG International Survey of Corporate Responsibility Reporting 2005.

[81] The 2007 Corporate Climate Communications Study, available online at http://www.corporateregister.com.

[82] "Companies Increasingly Report On Sustainability Issues: Investors Hampered by Lack of Quality, Compatibility," *Social Investment Forum*, news release, November 19, 2007.

Chapter Ten Communicate Your Carbon War Strategies: Companies Post their Climate Change Positions

A central influence in this process—and the object of "appeasement" criticism from some in the business community—is the Global Reporting Initiative (GRI), the entity launched in 1997 by the United Nations Environment Programme and Ceres, the investor-focused action group that had previously sought company commitments and information with its *Valdez* principles. GRI began its Sustainability Reporting Guidelines in 2000. Part of the acceptance of these guidelines by the business community was the fact that they built on a concept that has respect among many companies: "the triple bottom line" of economic, environmental and social performance.[83] The growing viability of socially responsible investment funds convinces many business leaders that a company's commitment to sustainability is a plus with investors as well as with customers and potential employees. A study in 2006 found that 75 percent of the Fortune Global 250 companies who produced sustainability reports cited economic reasons for doing so.[84] GRI's push for commitments, sanctioned by the involvement of so many companies such as GE, will likely result in increased data collection in various business sectors—for example, food processing—enabling GRI to set up sector-specific performance indicators.[85]

More companies are adhering to GRI standards as they gain respect, now in their third iteration.[86] Some companies—such as Gap and Nike—have even included the GRI Index itself in their reports, as a guide to their content.

Voluntary transparency is expressed in a number of other reporting mechanisms. As one leading example, IBM—which has responded to the Carbon Disclosure Project since CDP began in 2002—was one of three manufacturers to begin reporting its greenhouse gas emissions under the U.S. Department of Energy's Voluntary Greenhouse Gas Emissions Reporting in 1995, and it continues to provide its data.

IBM was also one of the early participants in the emissions trading mechanisms set up by the Chicago Climate Exchange. This puts a company's direct and indirect CO_2 emissions from operations on the

---

[83]"About GRI," http://www.globalreporting.org.

[84]"Carrots and Sticks for Starters," The United Nations Environmental Programme and KMPG's Global Sustainability Services, 2006.

[85]"Food companies take responsibility for improved sustainability reporting," GRI, news release, November 20, 2007.

[86]"About GRI," http://www.globalreporting.org.

record, and the reports are verified by the National Association of Securities Dealers.

Companies are vulnerable to criticism if information and claims in their reports are not verified. Large multinationals have led the way in this area: of the 335 companies in the Global FT500 who issued a citizenship or sustainability report in 2007, 44 percent included an external, third-party verification of the reports' statements and data. By comparison, of the 2,500 companies within CorporateRegister.com's broader database of smaller companies that issued reports, only 27 percent included independent third-party assurance.[87] Increasing emphasis on third-party verification centers on independent groups such as Calvert Social Research and the Dow Jones Sustainability Index/SAM.

VOLUNTARY REPORTING INITIATIVES
(FROM IBM WEB SITE)

In response to growing concerns about global climate change, Congress authorized the Voluntary Reporting of Greenhouse Gases Program under Section 1605(b) of the Energy Policy Act of 1992. Administered by the U.S. Department of Energy's (DOE) Energy Information Administration, this program is an ideal means for "Climate Wise Partners" to track and be recognized for their emissions reduction achievements. IBM is one of the three industrial companies (GM and Johnson & Johnson are the other two) that started in the first reporting year 1995. Through this program, each year, IBM reports data on CO_2 emissions (both direct and indirect) and CO_2 reductions from all its U.S. operating facilities including manufacturing and development as well as non-manufacturing sales and distribution locations. In October 2000, the U.S. Environmental Protection Agency fully integrated the Climate Wise and ENERGY STAR® Buildings programs into one ENERGY STAR program to provide all partners better access to energy star tools and technical resources making it easier to be competitive through strategic energy management and to protect the environment.

As per the 2003 Voluntary Reporting of Greenhouse Gases report filed with the DOE, IBM's estimated total CO_2 emissions associated with energy consumption from all U.S. facilities were 1.72 million U.S. (short) tons. From these U.S. facilities, IBM avoided over 122,000 tons of CO_2 emissions due to energy conservation projects.

---

[87]The 2007 Corporate Climate Communications Study, available online at http://www.corporateregister.com.

Chapter Ten Communicate Your Carbon War Strategies:
Companies Post their Climate Change Positions

Communicate to Create Stakeholders

Why does a company communicate? *How* are companies—and especially American companies who work under the distinctive carbon war conditions of American democracy—communicating their climate change and sustainability strategies?

My experience convinces me that the goal of corporate communication is to assist, and in many respects to lead, in the creation of stakeholders who are needed to affirm management's right to compete successfully in the marketplace. Create stakeholders, beat competitors—to my way of thinking, after observing the work of a lot of successful public relations people, that is ongoing job one. It is the job that wraps around every aspect of enterprise, that *engages* the chief communications officer with all others in top management and to add value in decision making and execution. As to *how* this key-to-success is put to work: *as fully, consistently, honestly, transparently and effectively as possible to lock in with stakeholders and therefore go most of the way toward accomplishing ongoing job one.*

To raise stakeholder trust, turn around critics and beat the competition, communication of the company's climate change and sustainability strategies will use proven guidelines on sustainability Web sites, in print and in presentations by company spokespersons:

   Tell the company story in terms that each stakeholder group will understand;
   Tell the whole story, the good and the bad; and
   Prove your commitments and claims with data, examples and verifications.

The essential questions to be answered are: *Does this show that we understand each stakeholder's interest and expectation? Does this prove that we are addressing, relevant to our mission, the social and economic (and probably, through public policy positions, political) issues that come within our accountability to our business and our stakeholders?*

Web is the Way

The Internet is the essential channel for your company's sustainability report to reach your company's stakeholders. As counselors like Jim Sloan point out, Web versions of sustainability reports—which in earlier days consisted of PDFs of the print version—have become flexible and adaptable, lending themselves to the multilayered presentation of valuable information, and enabling updating and changes. A high-level summary, presented on a site's main page, is adequate for readers to scan the company's major commitments and broad involvements. Most sites then let the visitor drill down for detailed data about specific activities, emissions, projects and plans. Advanced sites allow two-way communication—a move toward enhancing the company's connection to the tools and techniques of social media.

Corporate sustainability communication has a special advantage now, in Greening 2.0. Unlike much of the early greening period when companies struggled to get messages out, before the advanced features and popularity of the Internet, communicating now rests largely on what moves and what lives on the Web. To find out how your competitors (and business partners) are telling their version of the carbon war story, open their electronic front doors.

Climate change and sustainability are occupying huge segments of Web space. In the sociopolitical spheres that are the focus of this book, "climate change" or "global warming" are ranked near the top of daily topic searches. Small wonder, then, companies with something to say about climate change and carbon war strategies are making it easy for Internet searchers to come in and inspect their premises. Companies use their Web sites to stake claims, explain positions, reveal information (sometimes giving access to entire "environmental" or "sustainability" reports, with technical data on carbon and other releases), and highlight their climate-friendly business entries of products and services.

General Motors pioneered online carbon-war–climate-change dialogue for carmakers and their critics and stakeholders in 2008, with its GMnext site which boldly began with an executive's blog publicly musing on whether the site fits the charge of corporate greenwashing.[88]

---

[88] According to a *Financial Times* article on 2/7/08, shortly after the site was launched, "an assault by environmental activists on General Motors' 100th-anniversary Web site has turned into a pioneering online chat between the carmaker and its critic . . . GM's vice-president for North American field sales and service spent an hour yesterday fielding questions from environmental campaigners and others on the subject of 'corporate greenwashing'; posts included pictures of protestors at the Detroit motor show calling on the industry to combat climate change and to create more environmentally friendly jobs. A GM spokesperson said the site was another element in corporate transparency. 'We want to get as many voices in this debate as possible,' he told FT. 'We can't just pick the friendly questions if we want this to be a credible conversation.'"

Chapter Ten Communicate Your Carbon War Strategies: Companies Post their Climate Change Positions

CARBON WAR STRATEGIES

The war on carbon—the hugely popular fight to save humanity from uncommon climate change—requires companies to adjust their strategies for continued business success in heretofore unchartered realms of social accountability. Government action, backed by popular opinion as well as being encouraged by some companies' business initiatives, is beginning to forever change the American economy's carbon-endowed foundation. Companies and their representatives in Congress, as well as at least 15 state legislatures led by California, are examining targets, costs and options on constraining carbon emissions from all sources. The effects on every aspect of business—from factories and power plants to insurance and the Internet—will require new calculations of costs and risks, and new strategies for widened opportunities for products, services and lines of business.

The primary questions raised in this book are: what is your company's climate change and sustainability strategy in these war conditions, and how will you execute and communicate about it? If you're in the C-suite of any large company, here are three strategies that can be considered for a rudimentary evaluation of how you are positioned and from that, how you will communicate in the years ahead.

PROFIT IN ENERGY EFFICIENCY

Cuts in greenhouse gas emissions to make the world safe from global warming can be achieved at a net profit to the global economy, according to a 2008 study which concludes that investment in energy efficiency of about $170 billion a year worldwide would yield a profit of about 17 percent or $29 billion.

Mindy Lubber, president of Ceres, a coalition of institutional investors that commissioned the report, said: "This is the message financial leaders need to hear: there is huge opportunity [in energy efficiency] and those moving money around are going to make the difference in this. Efficiency is the fastest, cheapest way to reduce greenhouse gases and could bring large profits to the global economy." Lubber said institutional investors are seeking more energy-efficient property portfolios.

Source: "Study finds profit in cutting emissions" by Fiona Harvey, Financial Times *February 14, 2008*

Carbon War Strategy A: Get With the Program. This is the advocacy strategy. The company reads the public policy trend, accepts it and decides to make it a friend. Executives get involved with policy makers to shape government rules toward win-win outcomes. If this were a card game, the bet would be to take some share of the pot. Onlookers—or company stakeholders—are told how this is in their interest.

Carbon War Strategy B: Fight the Program. Call this the contrarian strategy. The climate change policy trend-line is so alien to the company that it looks like a noose. Executives turn to their political representatives and policy makers to slow walk or stop indicated outcomes that will have a hugely adverse effect on the company and its business. The bet is on playing to break up the game. Stakeholders are frightened by the dire prospects and the communications from the company are toward the valiant defense of their interests, seeking support for political resistance.

Carbon War Strategy C: Watch the Program. The straddle strategy. Public policy trending seems fairly clear but management asks, *why show our cards just yet?* Executives keep a low profile, maintain relations with policy makers and watch how the advocates and contrarians play. The idea is to win a late pot, maybe with a surprise play. Stakeholders are assured that management is aware of the changed competitive environment; general commitments are expressed toward social and economic accountabilities.

GE, DuPont, Duke Energy and Caterpillar were among the early *Strategy A* adopters. Strategizing their respective green-to-gold business prospects, they began working with policy leaders in Washington and key states, to help take charge of social, economic and political outcomes. By the same token, General Motors was the first among the highly-targeted auto industry to move toward an economy-wide approach to carbon dioxide control, getting with the program that ultimately resulted in a government-control outcome more favorable on most counts than proposed alternatives.

As for *Strategy B?* Resistance is risky. While companies may stand on principle, and support thoughtful counterviews from think tanks and industry associations, few companies can afford to outright stand in the road and face the steamroller. The exception may be the fiercely successful, proudly independent, private, family-owned organization that has long produced one carbon-heavy product and has little or no market-shift flexibility. An example of such a stalwart appeared early in the greenhouse gas issue, when an independent coal mining firm's

Chapter Ten Communicate Your Carbon War Strategies:
Companies Post their Climate Change Positions

CEO put up a brave, basically logical and certainly pugnacious battle in appearances before Congress and in the media.[89] The fight-the-system strategy was not supported by others in the industry.

Strategy C has been and may continue to be the most populated area as the storm clouds gather over American business, and as companies wait for financially sunny opportunities. The wait-and-C-suite straddle generates a communication strategy characterized by statements of principles, go-slower ideas, and evidence of carbon-footprint-reducing moves.

ONLINE REVIEW OF COMPANY COMMUNICATION

My associates in EnviroComm helped me prepare to write this book by scanning the Web for insights on how companies are telling their climate change and sustainability stories. This informal survey had two goals: find corporate communications that are good examples, if not benchmarks, that chief communications officers and their corporate teams can consider. We settled on sorting for corporate sites in which, as of the spring of 2008, companies either took a forward and active position (advocates of Carbon War Strategy A: *Get With the Program*) or expressed some level of acknowledgment or commitment (our cautionary category Strategy C: *Watch the Program* stance).

To get useful insights on corporate climate-change communication, the team read relevant statements posted on 200 company sites, roughly the larger Fortune 500 companies. We checked home pages and went within easy-access site sections, using tabs or other devices on the home page or using available search devices for key words including sustainability, environment, energy efficiency, global warming and climate change.

Of the sites reviewed, the more aggressive communications were from companies associated with coalitions and groups with forward positions on global warming response. These generally accepted the urgent (or "crisis") climate change situation and the need for collaborative action engaging public and private interests. Some took very specific positions on legislative approaches. For instance, at the beginning of our survey, BP, Dow, PG&E, GE, Duke and Shell had joined the U.S. Climate Action Partnership,[90]

---

[89] Kimberley Strassel's "Coal Man" interview, in the *Wall Street Journal* of May 19–20, 2007, is instructive for corporate communicators who may find themselves fighting against the odds on a social business issue.

[90] http://www.us-cap.org.

whose members from both the NGO and corporate communities advocate federal legislation on cap-and-trade and other control mandates. We found also a more aggressive and public-positive set of messages from companies active in trade association, government and general business programs. For example, among the more than 120 companies listed at the time of our review as part of the U. S. Environmental Protection Agency's Climate Leaders program companies including Abbott, Citigroup, Dell, Pfizer, and United Technologies were rather clearly Strategy A: *Get With the Program* communicators.[91] We found many of these same companies among the more than 125 members of the Business Roundtable's RESOLVE[92] program (Responsible Environmental Steps, Opportunities to Lead by Voluntary Efforts). And there was a general scattering of Strategy A enterprises in various industry groups. (The climate change policy positions taken by several associations, think tanks and other groups appear in the Appendix.)

What about companies falling into the extremely limited category Strategy B: *Fight the Program* posture? As brave as these windmill tilters may be, they are generally less communicative, less accessible and less apt to rationalize social concern and business case for global warming, and therefore less useful to communicators seeking benchmark examples. No such company is provided in our list.

Inferences for Carbon War Communicators

Notes on our review of company climate change and sustainability positions appear in the *Guide* that begins on page 177 of this book. Executives in any C-suite will find a fairly wide range of approaches, in both style and content. Our team drew these inferences for the chief communications officer:

- *Take charge of your language.* This is a communications tactic that starts by being able to use the Peter Drucker-type orientation: *Who is your stakeholder? What does he want (and understand)?* Therefore, what do you do? If you're clear on that, the content, style and language on the Web site should be much easier and most effective. In our survey, "climate change" and not "global

---

[91] http://www.epa.gov.

[92] http://www.businessroundtable.org.

Chapter Ten Communicate Your Carbon War Strategies:
Companies Post their Climate Change Positions

warming" seemed to be the more comfortable semantic haven, but we see the political loadings of "global warming" diminishing through common use and the term may well be the best way to connect a company with its stakeholders. "Sustainability" is popular, with some evidence of "corporate sustainability" and I at least am biased toward that since I think it wraps around Greening 2.0 to include energy—and engages the power of sociopolitical interaction.

- *Show sociopolitical interaction.* We saw the tripartite accountabilities—social, economic and political—showing up on effective sites. The critical tenet of stakeholder communications—make it clear that you understand and care about the stakeholders' interest or concern—was easy to grasp when the site provided examples (text or photographs) of the company engaged with stakeholders including local groups, business partners, regulators and legislators.
- *Make the business connection.* Examples of products, services and business lines benefiting from a smart, sensitive carbon-war strategy were shown by some companies. But we didn't see what we consider to be enough connection to corporate economics or financials in many of the green or climate-change sections. Care for the investor, as well as for business partners and employees, need not be overwhelmed by the softer, social side. At minimum, links or tags to "investor information" need to be prominent on the green site.
- *Prove it.* As open doors to show evidence and build trust, the best examples were sites that made green progress and especially carbon-reduction targets clear and trackable. Commitments were best expressed in incremental terms (a specific number by a specific date, with milestones to get there), with graphs, tables and comment as to how progress will be achieved. The proof clinchers were respectable third-party validations and endorsements.
- *Keep it fresh.* We found some old content on company sites that conflict with current information—effectively undercutting corporate communications and those who convey it in other forums. Best case seems to be to abandon a lot of the archived information, to update the site constantly, to keep the graphics

and all the tabs and keys and functions working. At the same time, know that when you take down something you've posted, somebody out there in the news or watchdog community will call you on it and try to make hay with it.

Finally, while we didn't find much of it on many sites, feedback needs to be facilitated through various devices for contacting the company or participating on an affiliated blog. This carries risk but nothing that is unmanageable. It's your site and you control it, so solicit and prepare for feedback. Think through what you're posting. Have talking points and fact sheets available to your executive team and employees. Plan your handling of questions, criticisms, blog postings and additional communication opportunities created by opening your electronic doors. Bottom line: *The Internet is the company's primary carbon-war communications channel for building trust and corporate sustainability.*

Chapter Ten Communicate Your Carbon War Strategies:
Companies Post their Climate Change Positions

WEB SITES TO VIEW FOR COMPANY RANKINGS

The Carbon Disclosure Project, operating in London with support from international investment companies and the U. S. Environmental Protection Agency, among many others, has asked about 2,500 of the world's largest companies to respond publicly to a series of questionnaires about their carbon-related activities. About half of them have complied, and you can look at the results at www.cdproject.net. ClimateCounts at www.climatecounts.org, funded by Stonyfield Farms and launched in collaboration with Clean Air-Cool Planet (www.cleanair-coolplanet.org), visits company Web sites for information used to rank selected firms on scorecards. While the rankings are somewhat arbitrary, the site has received a lot of publicity and is typical of what companies can expect from corporate watch channels.

Other sites useful for comments on business greening are Joel Makower's site, www.greenbiz.com, and the well-written but not necessarily business-friendly Grist.

"Do you really dare put your head above the parapet by touting your greenness and attract very knowledgeable consumers who are going to crawl all over your business…? If consumers … can catch you telling a half-truth, they will."

—*Mike Longhurst, senior VP at McCann Erickson, London* [93]

"There were ups and downs in forging this partnership and there were definitely moments when it wasn't clear if we could all agree on how to proceed. And at these moments, it was the CEOs who stood up and insisted on success, and that's really inspiring."

—*David Yarnold, Environmental Defense Fund* [94]

---

[93] Quoted in a 2007 *AdAge* article.

[94] From an online interview about the formation of the U.S. Climate Action Partnership, accessed February 29, 2008.

Chapter Eleven

Communication Comeback:
From Greenwashing Defense to Collaborative Offense

In the early years of greening, when company communicators were ducking eco-flak and struggling to explain their environmental responsibility, a powerful one-word, killer put-down was invented. It came from the environmental activist community. It was aimed at business. The word was "greenwashing." Used to mock business green talk, it meant the company was painting over bad stuff with good words. Used to combat corporate communicators, it meant we were not to be trusted when we reported green commitments or progress.

Simple, understandable and sticky, the epithet "greenwashing" was a conversation stopper. And personally painful, if you happened to be on the receiving end—which brings me to a personal account …

THE DAY I BOMBED IN BOSTON

Having come out of the mining and chemicals business and gone into counseling manufacturers, I had been targeted by pioneer environmental crusaders for my role in communicating industry's plans and achievements. Ultimately the chemical industry's efforts in which I was prominently involved would lead to the *Responsible Care* program with industry/community initiatives. The program was aimed at setting a rational pace for green success, in partnership with communities and environmental groups. It was progressive and we talked about it, but corporate green talk was in those days seen by some in the environmental community as a flare summoning activists to attack. Thus I came to face formal greenwash charges, as the lone business representative on a panel at a Society of Environmental Journalists conference.

Greenies ruled the room that wintry day in Boston and the mood was surly—or so it seemed to me as I sat at the speakers table waiting my turn.

The moderator of the panel had written a corporate watchdog book that described me as a rising purveyor of greenwashing for nasty industries. I had surprised him, I think, by placing a phone call to him after the book came out and volunteering to be a speaker on the conference panel. And this was the fated day. I followed three speakers. There were the author's opening comments, not too bad really—he didn't belabor the evil with which he had associated me in his book. Next came a searing indictment of business from a leading environmental group spokesman. He was followed by a university researcher with results of a survey on corporate environmental responsibility—decidedly unflattering. But who could argue with a survey? I certainly wasn't prepared to—and it was my turn to speak.

My message (quoting from the script now yellowing in my files) went toward the practical. "You're interested; we're interested in solving environmental problems. You may be ahead of us in your view of the destination, but we're heading in the same direction and you need us to get there." Going for collaboration, I tried out the phrase we would later develop (and use at the upcoming 1992 UN Earth Summit): "Companies are not asking you to trust them but to track them as they move down the road. We need to keep talking . . . neither of us should try to shut down the dialogue. . . ."

The hostile, head-shaking chill in the room was numbing—nobody wanted to hear about company and NGO reps linking up and skipping down the lane together. I think I got one question that I could answer and I got a three-minute lecture from an audience member who assured me that I was out of touch with reality, and, in effect, either a capitalist tool or a PR fool or possibly both. I was done. The moderator interrupted the silence with his thanks for my courage in coming out. A colleague who went with me to Boston from Washington suggested we head for the airport while we still had our scalps.

"Greenwashing" Power to Clog Dialogue

The negative jibe of "greenwashing" survives as a device to interrupt corporate dialogue with its stakeholders, and at the end of this chapter, I will comment on that. Here, however, I would like to offer to the green rethinkers in the C-suite my evidence that the "greenwashing" epithet has in the carbon-war era lost much of its going-green-era punch. The evidence is this: much greater alignment of corporate PR with marketing/

Chapter Eleven Communication Comeback: From Greenwashing Defense to Collaborative Offense

advertising, the substantial rise of corporate transparency on green as well as financial matters, common use by companies of third party validations, and the very interesting prospect of shared pie—both social and economic return—when companies and NGOs talk seriously with one another about cutting carbon.

(1) Alignment of PR and advertising. In about the same timeframe as my interesting engagement with environmental journalists in Boston, American companies got an emphatic lesson that positive green talk can backlash if PR and advertising are out of synch. Soon after the chemical company DuPont made a major decision to stop production of ozone-threatening chlorofluorocarbons and to take an out-front position in favor of the 1987 Montreal protocol to protect the stratosphere, DuPont's marketers got very creative. They put together a TV commercial with seals and penguins applauding the company's "going green" moves.

Rather than reassuring the company's stakeholders—and the politicians with whom the company had worked to give up its successful chemical product—the happy-creatures commercial had a reverse effect. It boosted anti-corporate activists and the power of greenwashing as a charge to put down good green works and dull the trust value of corporate messages.

Over the years, U.S. company communicators have moved away from extreme green exuberance, toward more measured, fact-filled reports on environment, health and safety. Publications and Web sites are heavier on the metrics that enable tracking, careful about claims and balanced. CCOs are building reputational value. While the ad folks in their creative approach to sell product benefits or corporate values in a compressed format may still tend to oversell—that is to say to go off the corporate message or get ahead of strategic corporate communications programs—there is considerably more alignment of the two disciplines as the market reality of the carbon war and the prospects for corporate sustainability are better understood among America's C-level executives.

Epitomizing the Greening 2.0 platforms of American corporations, and the apparent state of stakeholder approval, was the appearance of General Electric's "Ecomagination" commercial on television in 2005. The dancing elephant in a rain forest, with a head-bobbing fan club of tropical birds, touted the company's green social status without drawing greenwash charges from critics or objections from friendly stakeholders. It's tempting to suggest that the difference between this happy-animals commercial and DuPont's similar commercial two decades earlier simply shows how times, viewer tastes and tolerance have changed. That's true enough—but it also shows how corporate communications and advertising have aligned for the right kind of reaction from stakeholders.

GE's coordinated communications strategy combined the light, playful tone of the public commercial with a serious focus on financials. The ad was timed to the introduction of the company's high-dollar commitment to products that address energy needs and global warming issues. CEO Jeffrey Immelt personally took the Ecomagination message public in a number of personal appearances—from a prestigious National Press Club platform in Washington to talks with groups of investor analysts in New York. The investment community, conditioned to be skeptical of environmental investment, were told "green is green," and were given healthy sales and ROI projections from carbon-light products to back up the message.

(2) **Rise of corporate transparency.** The basis of any greenwashing charge is to suggest the company is covering up, hiding something. Two decades of corporate environmental disclosure have enabled some critics to advocate for more progress, but it has also taken sting away from charges of cover-up, at least with regard to manufacturing operational pollution. As far back as 1992 at the UN Earth Summit in Rio, companies knew the way to inoculate is to disclose. Companies organized the Business Council for Sustainable Development and announced its plan to measure operational pollution, plot controls and make data publicly available.

Voluntarily opening the kimono—to borrow the slightly shocking Oriental shorthand for disclosure—is the communicator's best offensive strategy to head off the discomfort of defense in today's carbon war. This communications situation is not that difficult, unless you get behind the curve of disclosure and commitment. The Carbon Disclosure Project, which originated in Europe and has gained international status, is one way to let critics know you're on the path toward carbon-related sustainability. The truth is, as I've learned from some communicators and preached to others, it is best to assume that there are no secrets, that everybody knows everything or will soon. With so many would-be critics out there looking for carbon-war malingerers, it's in the company's interest to take charge of the information outflow in a far more aggressive way. You need to make insiders out of as many outsiders as you can. It's not just Sarbanes Oxley. It's sociopolitical.

Chapter Eleven Communication Comeback:
From Greenwashing Defense to Collaborative Offense

(3) **Validation of respected third parties.** Green character witnesses provide two benefits to companies. It takes greenwashing off the table since experts are validating your green message. And it relieves the one-on-one tension between the company and an accusing green advocate by adding the third party. Third-party endorsements work at various levels.

Green operating standards can put companies into the winners' circle.

Guidelines from the International Standards Organization (ISO) and ASTM International have helped many companies move into sustainable environmental management. A company that can report its plants are ISO-14000 certified has opened the door for more respect from green groups, investors and other stakeholders, and has the added advantage of making employees part of the greening, building their pride in the company's social responsibility and—a plus for communications—adding their voice to public dialogue. The ISO standard provides specific guidelines for an environmental management system, and is recognized globally as a gold standard. ISO is a network of standards institutes in 157 countries, coordinated through a central secretariat in Geneva, Switzerland. In 2007, ISO issued its climate management standards that companies are now using to focus on CO_2, carbon restraint and energy efficiency in anticipation of requirements by government and company management. ISO is also an avenue into carbon standards that can prepare a company that wants to participate in the global carbon emissions market that is projected to reach $4 billion by 2012. Working with the International Emissions Trading Association, the Climate Group and the World Business Council on Sustainable Development, ISO has developed a standard to provide companies some level of assurance for the certification of voluntary offsets.

At the product level, marketing standards and guidelines are beginning to cover claims about global warming. Concern about climate change and marketers' moves to take advantage of it in various product offerings, the Federal Trade Commission has overhauled its voluntary Environmental Marketing Claims Guidelines which was last updated in pre-carbon-war 1998. The overhaul focused on the use in marketing of terms which were not under the 1998 guidelines, including "carbon offsets," "renewable energy," and "sustainable." Janice Podoll Frankle, an FTC Bureau of Consumer Protection attorney, explained in a

2007 issue of *Business Week*: "We want to make sure the guides reflect today's marketplace, consumer perceptions, and current science and technology."

Tools and performance criteria fit virtually every business.

Executives who are rethinking green strategies can find rules or guidelines for communication in almost every business sector. An example of impact is LEED. The Leadership in Energy and Environmental Design Green Building Rating System™ enables firms in architecture, construction and building management to qualify their sustainable building and development practices. Social activists as well as investors respect LEED, helping companies to get on the green side of public discussion with immunity to greenwash critics.

Recognitions from within a company's own circle—awards for a company that's on the board of the organization giving the award, for example—are less immuning to greenwash skepticism unless the selection is scrupulously substantiated through credentialled, objective judges, a feat most corporate communicators will recognize as difficult.

Government validations help a company raise a shield—if not win a green halo.

Check the Web sites of the U. S. Environmental Protection Agency, Department of Energy, Department of Defense, or any regulatory agency—including some state agencies, such as those in California—to see how your company could qualify for green or energy-efficiency listings. Voluntary programs like EPA's National Partnership for Environmental Priorities have removed more than 800,000 pounds of persistent chemicals from the environment since 2002. Companies in a companion program known as Performance Track are able to document progress in green management and energy use, achievement of measurable goals and transparency in public communication. EPA's Climate Leaders program qualifies companies in the carbon war.

The good news for CCOs is that government endorsement rules are put together with public input, so if you make the list you're qualifying on criteria at least partly approved by NGOs.

You may also be able to influence unauthorized ratings.

The "Climate Counts" scoreboard on the Web reflects the self-reported efforts of companies to address climate. If your company is

Chapter Eleven Communication Comeback:
From Greenwashing Defense to Collaborative Offense

not currently rated, take a look at their 22 criteria to see whether your company's climate footprint might match that of a winner's shoes. Review your communications to see how you compare with the Climate Counts evaluator to "publicly disclosing…climate actions clearly and comprehensively." The closer your green commitment conforms to those of the listed upright green players, then probably the more immune your company is to a hostile greenwash blast.

One useful guide in this regard is the landmark report on corporate sustainability, *State of Green Business 2008*, in which Joel Makower and the editors of GreenBiz.com introduced the GreenBiz Index, a set of 20 indicators of progress, tracking the resource use, emissions, and business practices of U.S. companies: carbon, materials, energy use, toxics intensity, clean-tech investments, e-waste recovery, paper use, and the climate impact of such factors as employee commuting. It is a virtual handbook on which to base modern, anti-greenwashing greenspeak.[95]

(4) Alignment of companies and NGOs. Two decades after my unnerving Boston appearance before NGOs who were skeptical of corporate greening, I still get a shiver recalling the way it was. However, that feeling is diminished by a small glow of satisfaction about how far we've come. By "we" I mean both sides, corporate and NGO.

At an Arthur W. Page Society conference on corporate responsibility in 2007[96], the dialogue between major companies' public relations officers and senior representatives of Environmental Defense and Greenpeace was rational and respectful. When they were pressed for tips on ways to add zing to a company's environmental communication, the visiting green leaders came up with what now must be considered a throw-back list—things that are under way and have been for some time, such as more environmental reports that show flaws and shortfalls as well as attainments, fewer glossy feel-good photos with kids and animals, and more collaboration with green groups. The long-trackable record of private-sector greening is accepted; and, much to my satisfaction, an academic speaker at the Page conference cited the substantial record of the chemical industry's Responsible Care community-focused communications program.

---

[95] http://www.stateofgreenbusiness.com.

[96] The conference was for corporate members, consultant agency heads and guests invited by the members, held in New York in April 2007.

Ongoing Transparency in Place

Green responsibility messages, stripped to their essentials, are today pretty much what we settled on after the 1992 come-to-Gaia UN Summit: *we care, we're committed to win-win outcomes, here's our evidence.* But now there's more to track, and the orderly march of compliance with escalating regulatory requirements has helped most companies climb the scale of green respectability.

Transparency and accountability—concepts that I suggested as communication necessities in my '92 Rio remarks—are routinely enabling stakeholder interaction. routinely enabling stakeholder interaction. Self-initiated reporting on emissions control has been joined by "disclosure" mechanisms originated by government and social action organizations, supported by participating companies. Corporate messages affirming greenism, with commitments and claims backed up by third-party evaluators, flow onto corporate Web sites. Corporate executive blogs are expanding the viability of green dialogue among companies and commentators.

The new ingredient for consistency for collaboration between companies and critics is the emphasis on climate change that binds energy to environment. In the current war on carbon, alliances are forming and firepower is coming together. At least three coalitions organized in Washington when Congress began getting serious about climate change action and an energy bill (spurred by Katrina, 2006 elections and Al Gore) put energy producers, consumers, green groups and government on the same side of the table in talks about carbon constraint, trading and alternatives to oil. The first green-friendly coal-power deal (TXU in Texas) got on the table, because private investors and NGOs came to terms on the threshold issue of future fuel source.[97]

Environmental Defense, which years ago shook hands with McDonald's on the deal to dump plastic clam-shell packaging to go cardboard (after environmentalists had first organized parents and kids to put the pressure on the company), is a prime mover toward mutual reward links. Corporate people at the Page conference learned that the current state of collaboration is so close that ED had opened an office in Bentonville, Arkansas to tie into Wal-Mart's global green-supplier

---

[97]More on the Texas utility deal, EDF, and other collaborators is provided in chapter five.

Chapter Eleven Communication Comeback:
From Greenwashing Defense to Collaborative Offense

influence. A conclave of green groups, FedEx and manufacturers to put battery power into diesel trucks was mentioned as one of many new moves in the transportation area, joining the acceleration of hybrid autos, supported by federal and state incentives, into the American marketplace. Prize-winning *San Jose Mercury News* former journalist David Yarnold, now a leading spokesman at Environmental Defense, has credited American CEO leadership as the deciding factor to pull together the U. S. Climate Action Partnership, the top coalition on legislative action on Capitol Hill at the start of the global warming debate. General Motors' launch of its climate-change-oriented Web site GMNext.com won respect from some in the green action community. "This is a great chance to speak directly to a major corporate executive," California-based Rain Forest Action Network commented. In its blog, the Network advised its members to join the dialogue opportunity opened by GM. A note to members said, "Bring your tough questions and your critical thinking skills."[98]

"Greenwash" Has Not Yet Dried Up

All, of course, is not exactly peachy keen between companies and green activists.

When General Motors made its bold communications transparency move in early 2008, calling for an interactive Web dialogue on climate change, some critics were eager to say the company's motive was "greenwashing." The company had challenges in managing what in early stages felt like an open door for enemy fire, but held its position and provided its stakeholders—and communication onlookers from other companies—a chance to see the future: frank, unfettered and ultimately the right move toward sustainable trust.

Consumer-level claims remain targets for greenwash accusation. Dozens of companies were listed in one well-publicized report by an environmental marketing firm as having committed one or more of the "six sins of greenwashing"[99] Communications on a range of products—from personal care to TVs to printers—were offered in evidence. TerraChoice Environmental Marketing, which said it used metrics from the Federal

---

[98] *Financial Times* article by Bernard Simon in Toronto, "GM invites green critics to join online debate," published February 7, 2008. Accessed online in late February 2008 at http://www.ft.com.

[99] The alleged "sins" are described in the Appendix.

Trade Commission (which has issued Environmental Marketing Claims Guidelines since 1992) and the Environmental Protection Agency, claimed that a majority of the environmental claims made for "more than 1000 products reviewed" were judged to be "either demonstrably false or risk misleading intended audiences." The report predictably touched off another wave of greenwashing charges, especially from eco-bloggers.

In Great Britain, the Advertising Standards Authority told two environmental agencies to alter an ad campaign after concluding it made unsubstantiated claims about the level of public opposition to airport expansion. The ad showed a picture of the Prime Minister and cited a poll claiming that "60% of the public want airport expansion to stop, because of its impact on climate change." The green groups were told not to repeat that statement or claims that "aviation accounts for 13% of UK CO_2 emissions—20% if you include return flights." The ads were deemed "misleading." And, to return to the Arthur W. Page conference in 2007, where environmental and corporate leaders were moving toward a sort of group hug on green issues, the Greenpeace spokesperson came forward as something of a porcupine. He conceded that his group's anti-corporate behavior is much less aggressive than in the old Greening 1.0 days when Greenpeacers invaded corporate property to hoist hostile banners—or worse. While generally inclined toward harmony, he said, he still gets a kick out of creating the odd corporate contretemps. And, indeed, in March of 2008, a measure of mischief was discerned at the kick-off of the Cottonelle Comfort Haven road show in New York City, when Greenpeace staffers handed leaflets to passers-by urging them not to buy toilet paper produced by Kimberly-Clark, charging the company with cutting down old-growth forests. My point, however, is that such inopportune occurrences are far fewer, somehow anachronistic in a time when so many are doing so much to take address green problems, and are much less news-worthy or worrisome to companies than ever before.

Carbonized Ground for Collaboration

As well, and I would say more important, if you're a corporate communicator you are well aware of the trap of greenwashing when positive green communication veers into chest-beating mode, when careful roll-outs fall victim to an ad agency's overzealous claims. And you are conditioned to dread and avoid if at all possible the worst case outcomes;

Chapter Eleven Communication Comeback:
From Greenwashing Defense to Collaborative Offense

when planned events and track records are overtaken by unexpected events: government action on a regulatory lapse, high-profile operational failures and the highly damaging management crisis. Somewhat similar to the case of Enron in the financial management category, BP has become the hapless exemplar of surprise and crisis in green management, with events confounding strategic reputation-building communications. The out-front social, green-leader positioning of BP, taken global behind an aggressive CEO through many years of focused public relations, marketing and advertising, was hollowed out by a serious of incidents that exposed apparent environmental and safety deficiencies. Greenwashing was not the worst of the negative charges that swept over the company, diminishing the company's green sunflower logo, and sending the communications function into both crisis communications and rethink modes, dealing with massive media and stakeholder issues, and plotting comeback strategies.

The point of this chapter remains: collaboration is in, forced largely by climate change conditions, and charges of greenwashing will have less viability. Corporate sustainability stands on the legs of management attention to financial performance, social accountability and political engagement. The pragmatic evolution of environmental and carbon economics has begun to open more common ground for collaboration with old and new stakeholders than has any previous set of factors. With global warming creating a broad, public focus on a common target, there is more opportunity for stakeholder and critic agreement on moves by companies in the direction of mutually agreeable green solutions. A great many more players have a stake in a successful outcome now than in the Greening 1.0 years in which the game was more zero-sum and companies more likely to come out loser. GE's "Ecomagination" platform, message and roll-out fit our description of corporate sustainability—creating strategies and communicating on the basis of agreement in the company's current social, economic and political contexts.

Congress and government forces at state and local levels are working to encourage markets for sustainable, low- or no-carbon products and services and to ease business' entry costs through incentives and grants. Companies and green activists who don't see mutual advantages in getting together in the war on carbon—which has begun to redefine much of what began as a strict and contentious focus on a series on many ecological targets—are very likely to lose reputation value and sociopolitical influence to those who do.

NAME CALLING DÉTENTE?

Name calling seems a little old fashioned—so Greening 1.0—when a green advocate who gets the full picture, as Environmental Defense's David Yarnold does, describes the common ground that greens found with companies in the U.S. Climate Action Partnership. Yarnold inferred that influential CEOs are putting together their own economic interests with evident social (he called it "moral") and political motivations, and thereby putting former antagonists on a practical path of accommodation. Concomitantly, in American C-suites where green rethinking is under way, I doubt that you could find very much interest in picking further fights with those who profess to represent social progress and I'm certain you would not find a single greenwash brush in the corporate communicator's pail. We're at the cusp of a name-calling détente.

There's one more reason why "greenwashing" charges no longer hold very much water, and that is that such charges can be flung both ways. As Joel Makower has asked, isn't there a greenwash tinge in some of the activist rhetoric—where big green talkers hope nobody mentions their big not-so-green actual lifestyles and their personal big carbon footprints? That is the state of the debate as it seems for now, and a comeback I didn't have that day I wasn't boffo in Boston.

Chapter Eleven Communication Comeback:
From Greenwashing Defense to Collaborative Offense

RED CHINA AND GREENWASHING

Commentary: October 2005—In the middle of a public-health scare in China about formaldehyde additives in domestic beer, an unusual commercial appeared on Chinese television. Cartoon lizards were shown partying in the wild, chugging beer and celebrating the health effects of beer produced by a particular brewery. According to the translation reported by U. S. news media, the reptiles' happy message about the sponsor's beer as they leaped and slid among the trees was: "No formaldehyde! It's healthy! It's cool!"

While the faint-praise message of "no formaldehyde!" might constitute a compelling benefit to market to Chinese beer drinkers, the commercial no doubt sent a shudder along the spines of American marketers and corporate communicators. In the U.S., green messages have to pass the red-face test of stakeholders and, if health risks are involved, the scrutiny of government regulators.

In the modern history of green communications, a practice under way in the U.S. since the mid-1960s, corporate communicators have generally learned how far to go in the direction of green halos self-bestowed on the company and its products. Green messages have to pass the red-face test, but we seem to have reached the place, in an evolving evaluation of the corporate right to talk about environmental commitment, where we can deliver the right message not only with a straight face, but also with a light, even fun, touch.

To return to the Chinese beer example, the question is not settled as to whether a business in a non-democratic society can achieve the same plane of confidence in social messaging.

While China impressively plows ahead on economic fronts, it exposes a soft underbelly of social neglect. Officials' vows to loosen strictures on public information are overtaken by events where any country is most sensitive: environmental and health episodes.

Going into the 2008 Olympic Games, serious issues of social accountability and clumsy, if not duplicitous communications cast a shadow on China's prime-time, world-stage readiness. Recognizing that communism (or, as I heard repeatedly from government officials during a tour with business people in 2005, "socialism with Chinese characteristics") does not sit comfortably with western ways, I suspect those who counsel China's leaders on green communications would benefit by slipping into the schoolroom of Western nations where for 30-plus years the tough lessons on these topics have been learned. Here are a few of them:

Lesson one: Trust follows performance.

Vowing transparency on social risks and government action, as President Hu of China has done on several occasions, is not the same as showing the evidence and inviting external evaluation. A chemical industry leader in the U.S., going onto the judging world stage at the original "Earth summit" held by the United Nations in Rio de Janeiro in 1992, hit the right note. "Track us, don't trust us," he said, as he and other Western countries put in place an audit system for measuring and public reporting on industry pollution.

Lesson two: Rules require stakeholders.

The American government builds on the advantage of democracy where, as Arthur Page once put it, public consent determines if an economic enterprise lives or dies. For every nation and every company, there are stakeholders – citizens, voters, employees, investors—who dare to expect social responsibility.

In a fiercely open society, there is give/take/resistance between private and public power, stakeholders get involved, state and congressional oversight is engaged and rules are hammered out. Successful companies in the West, who understand the interests of their stakeholders, pragmatically work with government to accept stretch clean-up targets. All the people aren't pleased all the time, but there's a regularized government process and the penalties imposed on violators are both financial and reputational.

Chapter Eleven Communication Comeback:
From Greenwashing Defense to Collaborative Offense

Lesson three: The bar keeps rising.

Clean air and water, waste management and public health protection are management givens. Since the 1970s, when the U.S. put in place a huge wave of environmental laws, a generation of adults has assumed a right to environmental and health protection, and greater assurances for the next generation. Climate change issues are pitched toward the future. Communication that succeeds in satisfying for now, will open the door for additional inquiry. The more that a company's stakeholders can access through technology and open information (Googling the Internet for hints, coalescing stakeholders into electronic communities), the more they will demand to know. The old rules of "right to know" are superseded by a growing demand to understand. There is a global quest for details, glimpses of the future, handles to the emerging issues. Clean-tech, nanotechnology, advanced pharmaceutical and medical doors are pushed open, and expectations don't look back.

It will be interesting to evaluate the success of environmental communications in China as this extraordinary economic power commits to some of the principles in the American democracy. It will be in the interests of all communicators if those in the world's greatest expanding economy can understand that management of constantly expanding realms of social capital is essential to economic success. And that communication aligned with performance is the true, honorable and lively art. And that the road to stakeholder hell is at least partly paved with chief executive visions and corporate intentions on green and other social issues.

"Business has always said, 'Give us certainty and we'll act.' And the government, instead, is giving uncertainty."
—***Michael Eckhart, President, American Council on Renewable Energy***[100]

"It has to become routine for any enterprise to ask at any change, even the most minor one, 'Who needs to be informed of this?'"
—***Peter F. Drucker***[101]

---

[100]Interview on E&ETV March 13, 2008.

[101]*Management Challenges for the 21st Century*, HarperCollins, NY 1999

Chapter Twelve

Corporate Greening 2.0: Green Walking into the Future

A single tear, trickling down the cheek of a Native American, was the most powerful and memorable message as greening got under way. In a captivating 1971 television commercial, the actor Iron Eyes Cody walks silently across a landscape littered with garbage. As he looks over the scene, the camera zooms to his solemn face, catching the tear as it rolls from his eye.[102]

The theme of that ad, timed to air on Earth Day, was easy to grasp.

America's former beauty, the stretches of untrammeled surfaces once known as the New World, is falling victim to commercialization and human abandon. Does anyone care? The tagline was: People start pollution, people can stop it.

It was about caring, as Phil Shabecoff noted in his book on environmentalism,[103] a theme well suited to the period's general social unrest, where the mood of the country was shifting with regard to socio-environmental issues. While the crying-Indian ad might have been typical of the fresh communications just budding in the 1970s' aggressive new environmental community, it was in fact placed by the Advertising Council and paid for by business members.

---

[102]Iron Eyes Cody, born April 3, 1907, in Louisiana as Espera DeCorti, is said by biographers not in fact to have been a Native American, but the son of two first-generation immigrants from Italy. Under the name he adopted as an actor, he presented himself as Native American, promoted Native American causes, and was honored by Hollywood's Native American community in 1995 with a recognition for his contributions. He died in 1999 in Los Angeles. Watch his classic commercial on YouTube: http://www.youtube.com/watch?v=X3QKEy0AIk.

[103]Phillip Shabecoff's book, *A Fierce Green Fire: The Environmental Movement*, first published in 1993 (now in paperback), is an authoritative resource on the birth and progression of greening and green organizations in America. After an accomplished career as journalist (the leading environmental reporter for the *New York Times* in the early days of greening, Phil moved into writing, editing, and publishing commercially. He founded and was executive publisher of the authoritative, daily and now Web-based *Greenwire*.

Critics blasted the campaign, calling it "a political decoy devised by corporate interests to divert public attention from the real issues of industrial waste" by placing responsibility for pollution on people rather than on companies.

Green messaging with roots in corporate communication had begun. And the Ad Council entry exposed the two-edged sword of corporate green talk. The commercial intended a message of corporate environmental empathy, but it served to energize the gathering attack on business and industry for its pollution. The test message *"we care"* came back *"oh yeah?"*

And so it has been, more or less, for the greater part of three decades. Going through the 1980s, companies were emboldened to take some credit for clean up and concern about the planet. Major examples were acid rain and the stratospheric ozone layer. Coal and cars came to grips with charges of sulfur dioxide and other pollutants drifting across breathable landscapes and communicated support of the monitoring and government's acid rain cap-and-trade accord that today is being cited as the best way to handle carbon emissions.

There was skepticism and backlash that bothered corporate communications strategists who were trying to do the right thing in that issue, but not like the trouble that one company—DuPont—faced a few years later in the ozone layer issue.

COMMUNICATIONS BACKLASH

A significant business in chemicals used in air conditioning and aerosols was under fire when scientists determined that emissions were eating holes in the ozone layer that protects the Earth from damaging ultraviolet radiation. After wrestling alongside other companies with what seemed to be unreasonable but relentless pressure to shut down the chemical (chlorofluorocarbon or CFC) business, DuPont managed to come up with alternatives acceptable to their customers—and led moves to phase out production.

There's where green communication got two-pronged and messy. On the one hand, corporate public relations people tried to be low key about their support of the 1987 landmark international agreement—known as the Montreal Protocol on Substances that Deplete the Ozone Layer—that called for DuPont and other companies to get out of the CFC and related businesses.

Chapter Twelve Corporate Greening 2.0:
Green Walking into the Future

At the same time this prudent message was overpowered by a highly creative TV commercial that celebrated DuPont's commitment by showing animals in the wild cheering for the company. Green activists—among whom were some particularly direct communicators who in the pre-Protocol days had put up a billboard in DuPont's headquarters city branding the CEO as an enemy of the Planet—jumped on the company for what they deemed to be a tasteless message looking for a green halo.

WALKING THE GREEN WALK

By the time I wrote my book on communicating environmental commitment—*Going Green*—in 1992, we generally understood that companies had to walk the green walk for a time, staying almost as quiet as the tearful Native American in that long-ago ad, before talking the green talk. And the talk had to follow performance. Companies had to think green, engage in green solutions and help government find green answers.

Communicators found green messages to be sticky—that is to say most effective—when they stated commitments, opened records for inspection, showed (sometimes surprising) innovations for green outcomes and invited stakeholders to join in these outcomes (including those that produce good financial outcomes), and largely left the praise to others.

When the talk gets ahead of the walk, companies have to back track. BP's escalating green messages were deflated by negative-news-making energy and environmental episodes—and a former creative consultant turned whistleblower, resulting in a painfully accusatory op-ed in the New York Times in 2006, illustrated with the company "sunflower" logo dripping with dirty oil. At Ford, a greens-pleasing promise of a large number of delivered hybrid cars proved unattainable, and had to be walked back by a chagrined CEO, giving fuel to eco-doubters and pressing the point that uneasy lies the corporate head that wears the green halo.

PROLOGUE TO GREENING 2.0

I continue to squirm a little when companies are walking the green walk so intently that they allow their marketers and advertising shops to tempt the double-edged sword of green talk with ads featuring vehicles or fuel that magically create clean air, or cartoon animals that dance in rainforests to brag about corporate ecology.

But times have changed, environment achievements by corporate America have been extraordinary and communication lessons have been learned. Leading companies, their communicators and their CEOs are earnest and open engagers in the action phase of the mother of all green issues, climate change.

A productive search for sustainability now forms the green theme—combining *"we care"* with *"yeah, we're showing how."* DuPont and Dow in the chemical business, Duke Energy and Entergy in the utility business, Wal-Mart in retail, Johnson & Johnson in pharmaceuticals, GE in energy, GM and Ford in autos, Caterpillar and Navistar in diesel, Shell and ConocoPhillips in oil, and many others are allying with government and former green critics to innovate, find carbon-constraint options and act to sustain both free enterprise and the common good. It's a far cry from the tearful Native American and the eco-activist backlash, and companies playing defense.

Now What?

As C-suite executives rethink their green strategies and plan corporate sustainability moves, what can be predicted that will affect corporate success?

Reaching for the future requires a good grasp on where you are and what you've got to work with. I like Peter Synge's advice that you need to keep one eye on the future goal and the other eye on current reality, adjusting as necessary so you stay on the course. That means both a hard look at the business mission and competition as they exist now, as well as best estimation of how the mission will be affected by competitive and resource forces in the future.

Projecting forward can reveal that today's position is risking tomorrow's competitive advantage. Stephen R. Covey, whose "habits of highly effective people" are useful to many corporate people, tells the story of the leader who climbs a tall tree to look at the trail ahead, only to shout back to his followers: "Wrong jungle!" In the carbon-war condition, corporate executives have to ask as they do as other risk factors develop, *Are we stuck in a disadvantageous course? Can we get from here to there?* Happily, they can ask as well, *What are our advantages? How can we get from here to there successfully?*

Jonathan Lash and Fred Wellington of the World Resources Institute recognized the implications of "companies that manage and mitigate

Chapter Twelve Corporate Greening 2.0: Green Walking into the Future

their exposure to climate-change risks while seeking new opportunities for profit will generate a competitive advantage over rivals" in a future in which carbon constraint will be a dominant factor. They cited early moves by Caterpillar, Wal-Mart and Goldman Sachs to create climate-change strategies, stating in the *Harvard Business Review*, "It's not enough (for a company's leaders) to do something; you have to do it better—and more quickly—than your competitors."[104] As laid out in chapter one, a fresh perspective on the company's operational, economic and financial factors will help top management deal with carbon war conditions.

Company business communications will need to reinforce the company's social or green accountability in its dialogue with critical stakeholders. CCOs must clearly understand that the carbon war has the momentum to keep green issues alive in new contexts, linked in new ways to essential corporate energy needs, and, at the same time, enlivening some positive prospects for new technologies, products, processes and markets. Management will need to stay close to what's happening in Congress, in the states and in other countries that impact company strategies. Marketing and sales will need to be sensitive to changing customer and critic attitudes toward green product claims. As a green marketing counselor has told her clients: A little preemptive caution is a small investment to maintain a strong—and trusted—brand.[105]

A Perception Edge for Business

It is clear that public expectations about solving climate change issues center on action by government and business. Public confidence that either or both will succeed is not so clear, but corporate sustainability seems to be gaining an edge. A global survey in early 2008 of senior sustainability practitioners—individuals connected to the practice and concept of sustainability in business, civil society, and academia—revealed a strong view that that critical issues related to climate change and sustainability were not being addressed adequately by either government or business leaders. However, the respondents scored the performance of business leaders much higher than that of elected officials. Over a third of the respondents rated the performance of government leaders as being very

---

[104]"Competitive Advantage of a Warming Planet," *Harvard Business Review*, March 2007.

[105]Wendy Jedlicka, president of Jedlicka Design Ltd. eco-packaging design firm; see http://www.jedlicka.com.

poor, while only 13 percent had this perception of business leaders. Asked to name company leaders, respondents most mentioned GE, Shell, Novo Nordisk, and Patagonia. The most frequently named "poor performer" in sustainability was ExxonMobil.

While CEOs and COOs might take some heart from public perception of being better at doing and communicating sustainable intentions, the group that did the global survey cautioned: "The jury is still out and awaiting more evidence on corporate performance. People with experience think business can and should lead, but they won't be bamboozled."[106] The bamboozle factor—skepticism on the receiving end of flowing information—is still going strong. When the sustainability pros were asked whether the current increase in communications about sustainability was "tactical and temporary" or "authentic (and) long-lasting," a mere 3 percent opted for "authentic." If you're in the C-suite, rethinking green challenges and positioning for sociopolitical success, how do you prove your enterprise is authentic?

AUTHENTIC ENTERPRISE COMMUNICATION

A survey of CEOs conducted by the Arthur W. Page Society[107] found corporate chiefs aware of their challenges as enterprise leaders as they encounter global economic change, empowerment of myriad new stakeholders and revolutionary communications driven by digital networks. These drivers of change, Page leaders observe, have created a business playing field where transparency, access to information and the ability to communicate are common public expectations.

To respond to—and, in fact, to take charge of—these new realities, companies will need management systems based on values that shape behavior and unite goals, extra effort to build stakeholder support across

---

[106] Jonathan J. Halperin, Director of Research & Advocacy in the Washington office of SustainAbility. The survey, conducted on-line by GlobeScan to identify drivers in and obstacles to the global sustainability agenda, drew 2,158 respondents from 20 countries. It was a centerpiece for discussion at the symposium, "What's Next for Sustainability," held at the Willard InterContinental Hotel in Washington DC in February 2008.

[107] The Page Society, incorporated in 1983, is a select membership organization for senior corporate and agency communication officers committed to the belief that executive-level public relations is central in the success of the corporation, and advancing principles drawn from the work of Arthur. W. Page, public relations officer for AT&T from 1927 to 1946. See the principles in the Appendix, and more information is available at awpagesociety.com.

Chapter Twelve Corporate Greening 2.0:
Green Walking into the Future

civil society at large and expert use of new-media tools. "The enterprise must consciously build and manage trust in all its dimensions," said the Page report on its survey. "Trust is no longer a function only of compliance with the law and business ethics.... It is a complex equation involving everything from employer/employee relations, financial management and corporate governance, to the quality of a company's goods and services and its responsiveness to societal issues."

The Page Society report, issued in 2008, said CEOs look to CCOs to take a more strategic and interactive role within the senior leadership of the company in driving changes. But Page underscored that "no single function will bear sole responsibility" for success in achieving what Page calls "the authentic enterprise," noting that all leaders in the enterprise will need to hone their collaborative skills.

WHAT IF ...?

Within the C-suite, rational sustainability decisions must be tested with rational what-if scenarios. This involves a collaborative examination of resources, markets, trade, customer and other factors—not the least being government, with climate-related and carbon-restraint technology moves tied to government rules and incentives. Congressional incentives for renewable energy are an example.

Supply-side economists have been reasonably successful in making the case to politicians that incentives to business—tax credits and tax cuts—not only get business into the arena to address social and economic problems. They can also pay for themselves. The rule of thumb—which held sway in the strong supply-side thinking of the 1980s and into the '90s—was that one dollar of tax credit generates four dollars of capital flow, sending revenue back to government as taxes are paid on economic growth. The problem is that tax cuts and incentives are not permanent nor necessarily reliable for business planning.

In 2008, as Congress considered whether to continue both investment and production tax credits for wind, solar, fuel cells and other power alternatives, leaders in those markets had to rethink business plans in the light of both overall recession-like conditions and the intentions on Capitol Hill. Most of the credits were due to expire in less than a year. Michael Eckhart, president of the American Council on Renewable Energy, told an interviewer in Washington that a company in the wind-farm business was like a student depending on a college scholarship that expires before she gets to school. "Why would you be investing in a wind farm development today that's not going to achieve closing until beyond when the tax credits expire?... You don't know what the rules of the road are. You don't know what to do. So, what it does is it shuts down the market until they wait for certainty. Business has always said, 'Give us certainty and we'll act.' And the government, instead, is giving uncertainty."[108] I have been in the corporate and consulting business communities for a long time. I built a business and ran it for 25 years. From this experience, and with the benefit of insight from corporate engagements, it's starkly obvious to me that the one thing certain about business is the persistence of uncertainty. Whether you're the CEO or the CFO, running a production unit or marketing products, in charge of IT or human resources or corporate communications—you know that forecasting is as risky as it is necessary. You need to be able to look ahead to get on the trail for competitive advantage. You need to consistently execute the mission. At the same time, you need to adjust to contexts as they become current. Pausing on the trail and climbing the tall tree to see if you're still in the jungle you thought you were in—that's a good thing to do. It helps you evaluate the elements of uncertainty.

As far as depending on government to help your business win in the carbon war? Considering all the options and timeframes for power alternatives, clean technology, carbon constraint in processes and products, all the risks of cost/price/competition carbonomics, plus the vagaries of politics—there is bound to be so much change that the best course for the company is to stay true to its strengths, develop and compete in markets that draw on these strengths, and use government incentives to advantage without becoming hooked on them. Peter Drucker warned corporate change leaders not to neglect the fundamentals of the enterprise: its mission, its values, its definition of performance and results. "Precisely because change is a constant, the foundations have to be extra strong," he said in *Management Challenges for the 21st Century*. Drucker was cautious about predicting the future, concluding that the only management policies likely to succeed are those that "make the future," sustaining and adapting current strengths. The successful sustainability leaders in American business will stay involved in sociopolitical agendas to minimize uncertain

---

[108] Interview on E&ETV March 13, 2008.

Chapter Twelve Corporate Greening 2.0:
Green Walking into the Future

outcomes affecting their own and their stakeholders' future interests. The potential for uncertainty will rise as climate change occupies executive and stakeholder thinking about the future, as will the risk of unfounded expectations.

CCO Role in Managing Expectations

Chief corporate communicators are positioned in most successful business firms to have some influence in the important matter of managing expectations. This influence is not to be taken lightly since it has the potential to keep executives on track—strategies that are achievable under current conditions—and to sustain and elevate the value of corporate communication in the executive team. The CCO must merit respect by bringing pertinent insights to C-suite thinking and, in the context of this book's topic, becoming part of the decision making on climate change and related Greening 2.0 matters. CCOs have the responsibility to be continually active in expectation management at two levels—C-suite and stakeholder.

Inside the Mind of the Stakeholder

This book offers a number of the questions worth raising as senior executives shape climate-change and sustainability strategies. In any sociopolitical matter there is almost always the risk of some level of misunderstanding among stakeholder groups about what the company plans to do next that will impact their respective interests. The question I have tried to keep in front of me in considering how to shape corporate communication is the question that the stakeholder is constantly asking: *"What's in it for me?"* As the considerable uncommon change occurs as the result of carbon-restraint and global-warming dynamics, stakeholders need to be able to understand the potential risks and rewards to them if they stay in your corner providing that blessing described decades ago by Arthur W. Page: permission for management to proceed. The value added by the CCO in creating stakeholders in corporate success will derive from understanding their respective interests and putting relevant interpretation on the table at the right time and at the highest effective levels inside the company.

Stakeholder perception is the CCO wheelhouse. CCOs know that stakeholder trust relies on fair deals and reliable information. The chief communicator's success in the C-suite is linked to the ability to get inside the mind of the stakeholder, understand the specific motivations that enable permission, and drive toward an honest answer to the "what's in it for me" question. The theme of this book is that the escalation of climate change as a sociopolitical concern will stimulate this question among stakeholders as it has rarely been stimulated before. And, boiling it down to the Hallmark card level—they don't care how much you know until they know how much you care—the response ideally will be that the company cares and here's the proof. Investors seek this affirmation with the borrowed clout of Sarbanes-Oxley. Employees look for it with personal and professional concern. Customers want to know what to expect of your offerings; suppliers, of your demands. And so on through the circles of participation in the company's path forward—which, its leaders hope, is sustainable and profitable.

Drucker noted that balancing change and continuity requires continuous work on information. Speaking of both internal and external information flow, Drucker said, "Information is particularly important when the change is not a mere improvement, but something truly new. It has to be a firm rule in any enterprise that wants to be successful as a change leader that there are *no surprises*."[109] Corporate communications will truly become sustainable communications when all the stakeholders are understood and their respective interests are addressed through an open, two-way flow of reliable information. That requires digging for input—from formal research to informal rambling in stakeholder circles—and forming relevant output—messages and actions that are true and trustworthy. Or as *Going Green* suggested in 1992 as the daily mantra for corporate PR readers: *Listen, understand, deliver.*

Chief communications officers serve several important functions in the C-suite. Among them is the tricky business of managing expectations—both at the stakeholder level and with the other executives inside the C-suite. The CCO assumes substantial accountability to control communications toward the just-right level of stakeholder reception.

---

[109] *Management Challenges for the 21st Century*, HarperCollins, NY, 1999.

Chapter Twelve Corporate Greening 2.0:
Green Walking into the Future

LASH-WELLINGTON CLIMATE CHANGE GUIDE FOR BUSINESS

In a conversation with Jonathan Lash at a sustainability conference in late 2007, I told him that the article that he and Fred Wellington (both of the World Resources Institute) wrote for *Harvard Business Review* was the single piece that I most often recommended to corporate executives rethinking their green strategies in the light of climate change. He cautioned that the situation is subject to factors, especially the nature of congressional and state government moves. Congress and the states move toward government requirements. Those factors are now developing and, if anything, the approach he and Fred outlined is more sound than ever. Following is an abstract from "Competitive Advantage on a Warming Planet", as it appeared in *HBR*, March 2007.

Whether you're in a traditional smokestack industry or a "clean" business like investment banking, your company will increasingly feel the effects of climate change. Executives typically manage environmental risk as a threefold problem of regulatory compliance, potential liability from industrial accidents, and pollutant release mitigation. But climate change presents business risks that are different in kind because the impact is global, the problem is long-term, and the harm is essentially irreversible. Furthermore, U.S. government policies have offered companies operating in the United States little guidance as to how environmental policy may change in the future. Ignoring the financial and competitive consequences of climate change could lead a company to formulate an inaccurate risk profile...

Climate change risks can be divided into six categories: regulatory (policies such as new emissions standards), products and technology (the development and marketing of climate-friendly products and services), litigation (lawsuits alleging environmental harm), reputational (how a company's environmental policies affect its brand), supply chain (potentially higher raw material and energy costs), and physical (such as an increase in the incidence of hurricanes).

The authors propose a four-step process for responding to climate change risk: Quantify your company's carbon footprint; identify the risks and opportunities you face; adapt your business in response; and do it better than your competitors.

"Al Gore is writing another book—and you can bet that climate change is shakin' in its boots ...

There are not yet plans to turn the sequel into a movie, but, nonetheless, we've already bought our tickets."

—*Grist* [110]

"The good news is we've got abundant technologies that are within our reach that we simply need to ramp up. If we can get the majority of the American people understanding that, [we] will probably let the folks inside Washington fight about the particulars ..."

—*Cathy Zoi, ACP* [111]

---

[110] The environmentlal blog in the popular Internet site, Grist.org, ran this in its September 7, 2007 entry.

[111] Cathy Zoi, CEO of the Alliance for Climate Protection, interviewed on E&ETV about former Vice President Gore's new line of commercials design to tilt public opinion on climate change, on April 1, 2008.

Chapter Thirteen

The Bridge to Sustainability:
What to Expect, What's Relevant?

Marilyn Laurie had never before put on a news conference. She had no way of knowing what this one would turn into. The young mother of two children, tired of sitting out the activism of the Sixties, Laurie simply knew that this cause—environmental cleanup—was right for her, and she knew enough about special events from her brief previous experience with advertising copy writing, that you needed to generate media with a news conference. She could not have imagined that on the day of the event, she would be on a platform facing a sea of people—estimated at a quarter of a million crowded into Union Square—standing between Paul Newman and New York Mayor John Lindsay, responding to dozens of reporters, from all the papers, TV and radio, *Time* and *Life* magazines. The date was April 22, 1970. Marilyn Laurie had helped New York City launch its first Earth Day. She could not have dreamed that the event she volunteered to publicize would make her the co-founder of an enduring international institution, and would put her on a personal career path as a corporate communications leader.

BIRTH OF EARTH DAY

In Washington DC, a year earlier, the idea of Earth Day had been conceived by a political patron in much the same way that John Adams and Thomas Jefferson had envisioned the first Independence Day—a broad, public display of commitment to a central, unifying idea. For the founding fathers, the idea was freedom from foreign rule. For U. S. Senator Gaylord Nelson, Democrat of Washington State, it was about the environment.

Nelson in 1969 had announced that Earth Day, beginning in 1970, would be a grassroots demonstration of public concern. It would condemn waste and pollution. Revering Rachel Carson's plea in *Silent Spring* seven years earlier, the observance would swell the common cause to save nature from human harm. With Nelson's congressional clout and with a prominent green activist, Denis Hayes, as the coordinator, Earth Day would thrust environment to a forward position in the national political and social agenda.

In a book of personal memoirs about the Sixties,[112] Laurie told the authors that her decision to get involved came on a Saturday morning in late 1969 while scanning the *Village Voice*. "The (environmental) movement was just exploding around us," Laurie recalls of her experience. "I just came in off the wall. I was looking for something to help with, and I found a cause. Or the cause found me."

Her husband Bob had brought the *Voice* paper home to show her the ad he had placed for his commercial art studio. She looked at his ad—and something else caught her eye. It was a small classified item about a public meeting to plan something called "Earth Day." Her thought was that here was something new that interested her—and that it was time for her to get out of the house and see if she could get involved. "I said to my husband, 'Look, this meeting is this afternoon. I'm going to go. Stay with the kids. I'm going to go.'"

Laurie learned a lesson about political activism. Fervor is not always associated with discipline. At least 400 people showed up for the advertised launch gathering. There was little order in the meeting and nothing accomplished except agreement to meet again the following week, which drew half the crowd as the first meeting. "Again, there was no plan, no real action," said Laurie, but the meetings continued. Finally, a handful—Laurie remembers only about five people—took over, volunteering to take on specific assignments, and staying with it until the job was done. Laurie agreed to handle the public relations. Her idea of a news conference was to invite every possible celebrity from show business, music and politics, in the hope some would say yes and the media would come to cover them. She was amazed when so many great people signed on, pledging to appear, to speak, to sing or make music. "Mayor Lindsay ... recognized that this was a better thing to be in front of than to be either behind or against," said Laurie, "and he responded accordingly."

---

[112]From *Camelot to Kent State: The Sixties Experience in the Words of Those Who Lived It*, 1987

Chapter Thirteen The Bridge to Sustainability: What to Expect, What's Relevant

Singers and actors—Pete Seeger, Paul Newman—showed up. To Laurie, it seemed that "everybody that we asked came"!

VOLUNTEER ACTIVIST TO DEDICATED EXECUTIVE

Laurie came to understand that, in her words, "leadership goes to those who are willing to go the distance"—and the former stay-at-home mom was about to enter a long professional journey. As a result of the success, which subsequently included working with the Mayor's Council on the Environment to encourage recycling, and creating a New York Times special supplement for Earth Day's first anniversary, Laurie was asked to head an environmental program at AT&T. Her involvement as a volunteer in what would become the nation's most active sociopolitical issue had put Laurie on a distinguished corporate career path. She would go on to discharge broad responsibilities as the first female senior vice president of AT&T, heading a 500-person global communications operation, subsequently becoming AT&T's Executive Vice President for Brand Strategy and Public Relations, chairman of the AT&T Grants Foundation, and a member of AT&T's 10-person Executive Committee.

Named as one of New York's "75 Most Influential Women," Laurie now heads a consulting firm on corporate public relations strategies including social responsibility.

CLIMATE CHANGE ENERGIZES EARTH DAY

As in New York, the inaugural event in 1970 was a huge success nationwide. Media reports from hundreds of cities and communities estimated that 20 million people, including 2,000 colleges and thousands of primary and secondary schools rallied for the first Earth Day. The Sixties' antiwar activism was fading, as the decade of the Seventies opened with public activism for environmental cleanup, drawing into it new commitments from citizens like Marilyn Laurie in Manhattan.

Today Earth Day is an institution embraced by volunteers, green activists, government and corporations worldwide. Its recent focus on climate change typifies the way in which the war on carbon has re-energized green activism, especially in the U.S.

In 2008, Earth Day was the biggest ever. "From Tokyo to Togo, to our flagship event on the National Mall in Washington and seven other

U.S. cities," reported the central organizing group called Earth Day Network, "we galvanized millions of people around the world behind a Call for Climate, our global warming action theme. Hundreds of millions of people from every corner of the planet raised their voices to urge significant and equitable action on climate change."[113]

The official U.S. government Web site, aimed especially at young people, keeps alive the vision laid down in 1969 by Senator Nelson and Denis Hayes with this statement: "Earth Day is a time to celebrate gains we have made and create new visions to accelerate environmental progress. Earth Day is a time to unite around new actions. Earth Day and every day is a time to act to protect our planet."[114] The "new actions" that lift the observance into relevance are impressive, embracing climate change, engaging a great many segments of American society, including business, and expanding globally. Hayes, who served as director of the National Renewable Energy Laboratory in President Jimmy Carter's administration, is credited with expanding the Earth Day Network to more than 180 nations.[115] He now chairs the board of the Network.

Nelson's original intention to raise the environmental issue on Washington's agenda was achieved in 2008. The Capitol telephone system handled thousands of calls placed by green activists asking Congress for a moratorium on new coal-fired plants, more investment in renewable energies, building efficiency and protection for the poor and the middle class in the transition to the new green economy. "If you haven't called for climate yet, do it right now!," the Earth Day Network said on its Web site, providing viewers with the central phone number for Congress, and a list of contact points in other countries.

In the streets near the Capitol, a rainstorm did not dampen the enthusiasm of environmental groups, citizens and Washington visitors peacefully demonstrating throughout the day and into the evening. Outside of the U.S., citizens called their governments and parliaments with the similar demands.

---

[113] See http://www.earthday.net/node/80.

[114] The site http://www.earthday.gov/govtsites.htm provides links to other government sites.

[115] Hayes is president of the Bullitt Foundation in Seattle, pressing to "make the Pacific Northwest the best-educated, most environmentally aware, most progressive corner of America—a global model for sustainable development." See Bullitt Foundation vision at http://www.bullitt.org/

Chapter Thirteen The Bridge to Sustainability: What to Expect, What's Relevant

COMPANIES AND EARTH DAY TODAY

In its more than three decades of commemoration, Earth Day has had a strong communications focus for corporations, especially in the U.S., signaling an annual need or opportunity for initiatives, news releases, internal communications and advertising promotion. This has been at times problematical. For at least several of the early years of the event, the business community was both the target for environmental blame as well as for financing various forms of Earth Day support.

In general, support from business today is a given, but is it any longer relevant?

Laurie, who now heads a strategic consulting firm, believes that it is, with the climate change connection as a significant factor. "As one who follows big trends," Laurie said, "I can't recall another issue developing with such astonishing rapidity, with such direct impact on most companies. It is up to each company to decide what that means, but certainly tying into the focus on climate change will make Earth Day a relevant opportunity for engaging with stakeholders."[116]

CROSS OVER THE BRIDGE

Here is my view on what American companies and their hard-working communicators can count on as the gate closes on the long, awesome and finally successful period of wrestling basic pollution into submission, and the sociopolitical beast of climate change stands challenge at the foot of the bridge to corporate sustainability.

1. Expect politicians to run hot and cold on global warming.
This sociopolitical issue will waver on the campaign trail. Opinion polls and media interest decide. During Al Gore's run for president in 2004, he fled from the issue when polls showed little vote value. The former vice president told a reporter: "Well, when I started talking about climate change you guys stopped writing."[117] In 2008, both Barack Obama and John McCain sparked media and moderate voter interest by inserting climate change in the campaign.

---

[116]Interview by the author, May 13, 2008.

[117]A recollection of the Gore-Bush campaign by John Fialka, editor of E&E's *ClimateWire*, interviewed by OnPoint host Monica Trauzzi on February 19, 2008.

2. ***Expect, however, government regulators in your life when the politicians move on.*** There is a disconnection between campaign trail rhetoric and federal or state legislative action. CEO Jeff Immelt told reporters that GE's long-range commitments and the effect of government rules will keep companies like his busy wrestling with business outcomes long after the media have "moved on to something else." For companies, prepare for climate change and sustainability politics to assume a long life in the forms of rules, standards, deadlines, compliance, reviews, inspections, lawsuits and general organizational burdens.

3. ***Expect customers to resist sacrifice and not to keep their green pledges.*** All the research shows that what people tell pollsters they want and will buy to save the planet may not necessarily be done when they look at the price. The "general public" wants something done, but by government and by companies, and at little or no extra charge or trouble to individuals.

4. ***Expect green probes by shareowners to go evergreen.*** ExxonMobil's dubious distinction as a resistant climate-change skeptic hit a high at the 2008 annual meeting when big shareowners (and Rockefeller family members) pushed punishing resolutions. When California State Teachers' Retirement System ($165 billion in assets) filed resolutions with other energy companies to set emissions-reduction targets, the fund's leader explained: "We view these things as long-term risks. It's a case of protecting our portfolio and maximizing our returns…(and) the fiduciary connection is pretty solid." Investors and insurers will zone in on your vulnerability even if somehow you've already accommodated the vocal green/global warming activist-media-political alliance.

5. ***Expect the possibility of drawing fire while doing good.*** Caterpillar's CEO had to explain to investors why the company signed on with advocates for cap-and-trade. Wal-Mart, clearly the world's most influential Corporate Greening 2.0 advocate, is called on the carpet by activists for products containing wood.

6. ***Expect somebody to be keeping score.*** A rising array of evaluators have found ways to weigh your company's climate change and sustainability value. Climate Counts' annually

Chapter Thirteen The Bridge to Sustainability: What to Expect, What's Relevant

updated scorecard reflects the self-reported efforts of companies to address climate change—or not. Leading companies bring focus—and force—to their efforts by signing on to criteria of credentialed evaluators such as the International Standards Organization and the Carbon Disclosure Project. Best move: be discovered to have done the right thing.

7. ***Expect to find some gold.*** Even politicians who have moved on will have left the incentive spigots as well as the rules running. Companies have turned operational environmental control into cost-saving efficiency. They've accepted government money to innovate in profitable technology and products. Green ability will seek economic viability. Merrill Lynch calls its climate change practice "Green & Gold" and points to a $70 billion carbon trades market. "This is a positive future, a future of abundance," said Fred Krupp of Environmental Defense after his survey of business conditions prior to writing his 2008 book on climate change. "This is a race that will create new billionaires. It's a huge opportunity for America."

8. ***Expect the unknowable.*** The known unknowns are the next promising and potentially threatening levels of nanotechnology, biotechnology, energy reliability and availability. New dimension ideas like energy from algae will touch off new opportunities and open new sociopolitical questions. What's the reliable forecast that you can count on? Well, you can:

9. ***Expect the enduring socio-politics of climate change.*** Al Gore's $300 million Alliance for Climate Protection campaign will stir the waters for years. Corporate Greening 2.0 has climbed atop an issue that can run forever. Activists, media and politicians can trade on it sustainably, and you will be drawn into the mix. Rep. John Dingell said in a Detroit Economic Club speech about global warming, "You will never be able to turn to your neighbor and say, 'Pardon me, but your end of the ship is sinking.' We are all in this boat together. We must all start paddling in the same direction." By the same token, there will be stakeholders, including former critics, who will want to become your collaborators in sustainable ventures. And so you, corporate C-suite player, can:

10. Expect to be into corporate sustainability—meaning it, making it rational to your company's distinctive stakeholders, working constantly to turn it into a three-way win (economic [think financial], social, political)—and vying with competitors who are trying to beat you at it.

RECONCILIATION AND CORPORATE COMMUNICATION

We have only just begun the task of reconciling the facets of climate-change impact on companies. The war on carbon will settle into some kind of endurable armistice. We will deal with the vestiges of Greening 1.0, the settlement of old cases and legacies. We will get used to Greening 2.0 with its black holes and golden rewards. Adaptation and innovation will be spawned and we will collaborate for common progress, if not survival. But C-suite management dynamics—executive planning and expectations with regard to climate change and sustainability strategies—will need to be continually reconciled with current and coming sociopolitical events.

If corporate public relations is, as I believe it is, the binding element in the management process of creating stakeholders in the company's success, the chief communications officer will face many differences needing reconciliation. Management actions planned and announced are not always executed. Trust and reputation are imperiled when company stakeholders expect more than is delivered. Media observes events and, depending on the outlet's bias, calls them differently. Some green groups attack a company's position while others ally with it. One investor's sweet spot is not the same as another's. And government, God bless it, often seems the most divided: ready to rise to the defense of economic contributors on the one hand, handicapping them on the other; flailing with sticks while offering carrots; and most egregiously, policy makers putting laws on the books that are endlessly interpreted as the resulting sea of regulations pours onto business, often leading to questions about the intent of Congress that can be worked through the legal system all the way to the Supreme Court. The high court settled the test of whether in fact carbon dioxide emissions are a pollutant to be regulated—the ruling back to the EPA was yes, it is—and that has led to the matter (apt never to be fully reconciled) of state versus federal rights to impose environmental and energy regulations.

Chapter Thirteen The Bridge to Sustainability: What to Expect, What's Relevant

TECHNOLOGY, SUSTAINABILITY AND CORPORATE COMMUNICATIONS

Predictions about the future for corporate communicators, including those made above, are fragile at best, but this one can be made with some confidence: CCOs can expect to have to look into both "bright and batty" sustainability ideas—and that's a good thing.

As the carbon war is popularized, you (as the senior information flow strategist) will be asked by officers and business unit people to look into news bits they've heard or read about. Nothing new about that. When inhouse emails or hallway conversations put these kinds of inquiries into your lap, you probably have a routine. You filter through what you know is relevant to the company's business, culture and stakeholders, and you or someone on your staff looks into it. There is always the possibility of an actual logical application, but your best achievable outcome is usually another learning experience.

However, let me offer an insight from *The Economist* to suggest a new wrinkle in what the future holds for companies wrestling with sustainability issues, a different sort of learning opportunity. That is an engagement with technocracy. The "bright and batty" phrase is borrowed from an *Economist* article about the surge of technology talent into the search for climate change and energy solutions. *The Economist* called this the new "supercool" frontier for successful dotcom era survivors. Restive in their state of early, often wealthy retirement, they are using their talent and money to mix it up with companies and investors on sustainability. The thrill of thinking outside the box with the possible prospect of new financial returns—"recapturing the good times of their youth," noted the Economist—has brought onto the field players such as Elon Musk, a cofounder of PayPal, who developed a battery-powered sports car; Larry Page and Sergey Brin, Google founders, whose venture called Google. org probes ways to make renewable energy cheaper than the use of coal; Vinod Khosla, a founder of Sun Microsystems, with sizable venture capital investments in renewable energy; Robert Metcalf, inventor of an ethernet system; also putting VC money behind sustainable enterprise; and others. Technocratic excitement is taking root in the form of energy and climate change notions, strategies and products that could flower into business blockbusters.

This has implications for C-suite communicators, whether in traditional or newer-age companies. First, you can expect the impact of advanced technology, as well as its talented and impatient advocates, on the options available in your company's business strategies. You will sense this impact not only inside the company as you receive clues or questions from peers or business people but also among stakeholders, starting with investors who will test the steadiness of management's hand in steering through the carbon-constrained business conditions. You will also discover the benefit of new areas of agreement—finding ways to plug directly into collaborations in the Technorati[118] communities, ranging from partnerships on advanced technology projects and announcements to joint ventures entered into by management.

Engagement at some level with respected technology developers and investors, including the newly influential players on the business landscape is a fresh opportunity in Corporate Greening 2.0. It will affect what, why and how you communicate with your company's current and prospective stakeholders.

We can draw a further, practical inference. Players from the dotcom decades will not only increasingly influence the business landscape. They will also be political players, influencing congressional as well as state legislative decisions that affect business. This is another significant post-dotcom, post-Greening 1.0, condition. While built-to-last companies were cleaning up pollution and negotiating a legislative and regulatory path with politicians at local, state and federal levels, the technology start-ups and larger enterprises trailblazing Internet and computer-use breakthroughs paid little attention to either greening or government. Washington DC, was a foreign field only mildly relevant. It was not until geeks-turned-executives like Bill Gates and enterprises grew big, led by Microsoft, that the technology community shook off isolation from the political community. They have gotten with the program. Microsoft showed the way, hiring public relations people and lobbyists in DC, getting to know government power leaders, and setting up a highly professional Political Action Committee that directs money to political campaigns nationally. Former dotcom activists, once fiercely independent, now work comfortably alongside representatives of built-to-last companies to help shape the laws and regulations that business will live with. Some are already working

---

[118]Recommended access to the nexus of sustainability and the high-technology business is http://technorati.com.

Chapter Thirteen The Bridge to Sustainability: What to Expect, What's Relevant

inside old-line companies, as well as in new-wave firms, as members of management, shaping and helping to implement strategies to innovate and to beat competitors in political, social and economic encounters. They are reaching for government support—research and development grants, contracts, favorable tweaking of legislation—that can help them succeed. In my view, this is just part of the process I have emphasized in this book—the strategic bonding, even if situational and temporary, of NGOs, public officials, corporate people, business partners, even competitors when their interests coincide and, basic to all this, interest-driven stakeholders, to address the uncommon condition of carbon constraint and innovation. The primary point here is that the climate change adaptation process is being refreshed and will be redefined by the participation of people knowledgeable in innovation and technology. Awareness of that reality will assist corporate sustainability strategies formed in any C-suite, and will provide productive new avenues to communicate with stakeholders.

The thread winding through these is the need for clarity, relevance and salient positions as the war on carbon affects business decision making. To manage, to win in the carbon-constrained years ahead, C-suite strategists will need a firm grasp on economic, social and political factors that are different from what they have been up to now. My emphasis on corporate communications derives from much evidence that an open, sharply focused and timely information flow between the company and its stakeholders is the foundation for understanding the factors that can be aligned for good, win-win and therefore sustainable outcomes. Corporate sustainability has become the testing ground of the grand, essential deal the company has with those who agree to support its business mission. In the era of Corporate Greening 2.0, CCOs will find ample opportunity to add value to the deal.

Guide to Corporate Climate Change and Sustainability Positions

"Make every business decision with sustainability in mind."
—David Kaplan, Chief Sustainability Officer, Dow Chemical[119]

"It's not enough (for a company's leaders) to do something; you have to do it better—and more quickly—than your competitors."
—Jonathan Lash and Fred Wellington[120]

"At Wal-Mart, the thought is if you've got this big [sustainability] department, then others will say sustainability is [that department's] job. If it's integrated into the business, everyone has a commitment to it."
—Kory Lundberg[121]

---

[119]From remarks at a Washington Sustainability Forum of the World Environment Center, March 25, 2008.

[120]World Resources Institute authors of "Competitive Advantage on a Warming Planet," *Harvard Business Review*, March 2007.

[121]Wal-Mart spokesperson, quoted in *Advertising Age* article, "Who's in Charge of Green?", June 9, 2008;http://adage.com.

CORPORATE CLIMATE CHANGE AND SUSTAINABILITY POSITIONS

*Drawn from Web Sites and Other Public Sources
During the Period July 2007–June 2008*

ABBOTT LABORATORIES

Abbott was the first *Fortune 500* company to commit to going "carbon neutral" with its U.S. fleet of vehicles by offering hybrid cars and other greener vehicles, and committing to purchase carbon credits to offset remaining vehicle emissions. Abbott's ongoing commitment to sustainable business practices led to the company being named for the past three years to the Dow Jones Sustainability World Index (DJSI World), which evaluates corporate economic, social and environmental performance.

As part of its participation in the EPA's Climate Leaders program, a voluntary industry-government partnership, Abbott has pledged to reduce its greenhouse gas emissions 10 percent by 2010 from 2004 levels, normalized by sales. The company is also participating in the **PHH GreenFleet** pilot program to improve the efficiency of fleet vehicles, thus reducing their emissions. The intent of the program sponsored by PHH Arval, a leading fleet management company, and the organization Environmental Defense, is to help companies make their fleet vehicles climate-neutral through conservation practices and GHG offsets.

Quotes from the Abbott site:

"Reducing our greenhouse gas emissions demonstrates our corporate-wide commitment to improving the environment and increasing the efficiency of our operations."
—Michael Warmuth, Vice President, Global Engineering Services.

"Abbott is demonstrating corporate climate change leadership by setting a long-term greenhouse gas reduction goal and committing to reduce its carbon footprint."
—Robert J. Meyers, Principal Deputy Assistant Administrator, EPA, Office of Air and Radiation.

Advanced Micro Devices, Inc.

Focused on meeting the needs of leading computing, wireless and consumer electronics companies to help them deliver high-performance, energy-efficient and visually realistic solutions, ADM has this to say (at www.amd.com) about climate change and energy efficiency:

"Global climate change presents a range of complex risks to the global community and our common vision of a prosperous future. AMD recognizes our responsibility as a global citizen to reduce our direct impacts on the environmental and to inspire and enable others to do the same. We do this through the development of energy efficient technology and the responsible design and operation of our facilities. Our commitment, strategy, and progress are reviewed annually at the executive level of the corporation."

After meeting a goal set in 2004 to reduce greenhouse gas emissions (as measured by kilogram carbon equivalent emissions/manufacturing index) 40 percent by 2007 against a baseline year of 2002, AMD set a new goal to cut normalized greenhouse gas emissions 33 percent by year end 2010 relative to the 2006 baseline. AMD's normalized greenhouse gas emissions goal is supported by two other goals: Reduce absolute perfluorocompound (PFC) emissions (metric tons of carbon equivalent emissions) by 50 percent by 2010 relative to the baseline year of 1995; and reduce *normalized energy use (as measured by kilowatt-hours/manufacturing index) by 40 percent by year end 2010 relative to the 2006 baseline.*

The company's Web site provides charts showing progress in reaching these goals and describes a number of programs including participation in Plug-In-Partners, a national initiative to demonstrate demand for hybrid electric vehicles.

American Electric Power

American Electric Power claims the largest fleet of diesel hybrid utility trucks in the country. The Hybrid Truck Users Forum estimates that the diesel engine and hybrid technology can save 1,000-1,500 gallons of fuel per utility truck annually (at 2008 diesel prices of $4 a gallon, that meant a savings of $4,000-$6,000 in fuel per truck annually); and annual greenhouse gas reductions of 11 to 16.5 tons of carbon dioxide are achieved per truck. With an order of 18 more trucks was announced in April 2008 of 18 hybrid trucks, AEP said it expected this addition to save $72,000-$108,000 in fuel costs and eliminate 200 to 300 tons of carbon dioxide emissions in a year's use.

AT&T

In 2005, AT&T became the first telecommunications company to join the California Climate Action Registry for operations in California. This registry, created by a California statute in 2001, helps companies and organizations track, publicly report and reduce emissions of the gases that can lead to global climate change. In addition, from 2001 to 2004, the company…decreased its annual gasoline-fuel consumption by 10 percent…decreased its annual diesel-fuel consumption by 20 percent. AT&T also recycles used oil and antifreeze from its fleet of service vehicles and backup power generators.

AT&T's technology provides the infrastructure for enabling environmental improvements throughout the global economy. Services such as teleconferencing, videoconferencing, electronic commerce, e-billing and NetMeeting are prime examples of AT&T technologies that directly reduce or eliminate energy and material consumption.

AT&T recognizes the need to use energy and other natural resources responsibly. Since 1998, AT&T has continued to drive cost-savings and improvements in energy efficiency across its facilities. AT&T's energy programs have improved the energy efficiency of data centers by nearly 25 percent.

BP

This international energy company—British Petroleum—posted these comments on its Web site in 2007: "In our operations since 2001 we have been aiming to offset, through energy efficiency projects, half of the underlying greenhouse gas (GHG) emissions that result from our growing business. After four years, we estimate that emissions growth of some 10 million tons has been offset by around five million tons of sustainable reductions. Many of 2005's efficiency gains were made possible by a five-year company-wide $350-million energy efficiency program that began in 2004. This program has enabled businesses to carry out sustainable energy-reducing activities, cutting costs as well as GHG emissions.

BP's position on greenhouse gas emissions:
- We support precautionary action on climate change although we recognize that aspects of the science are still the subject of expert debate and not fully proven
- We believe that climate change is a long-term issue, which needs to be tackled over the next fifty years or more. We support urgent but informed action to stabilize greenhouse gas (GHG) concentrations by achieving sustainable long-term emission reductions at the lowest possible cost
- There are many potential contributors to this goal. We support an inclusive approach that acknowledges the existence of many different starting points, perspectives, priorities and solutions
- We believe that governments and businesses must work together to create policy "spaces" in which economic progress, security of energy provision and emissions reductions can be achieved simultaneously. Such spaces can be defined by appropriate policy and regulation, while activity within them will be driven by market mechanisms
- We believe that the policy and regulatory interventions must support the development and implementation of appropriate technological solutions and also enable the amendment of market mechanisms as new knowledge around climate change emerges
- In our view, all energy sectors can help to address GHG emissions. With fossil fuels currently the source of 80% of the world's primary energy and likely to remain vital to global energy supply for at least 20 to 30 years, innovation to reduce carbon emissions from fossil fuels can make a major contribution to stabilization. Consequently, energy companies like ours have an important role to play in contributing to policy and education, enabling market mechanisms, developing and deploying new technological and commercial solutions based on both fossil fuel and new energy sources at large scale

With regard to engagement in public policy, BP said in 2007:
"We will focus our efforts on influencing policy, regulation and innovation. We can do this by working to develop inclusive, informed dialogue, supporting the creation of a 'level playing field' for low-carbon energy, encouraging market mechanisms and promoting

technological innovation and transparency; [and] continuing our business activities in energy efficiency, fuels switching, hydrogen power, solar energy, wind, biomass and natural sinks."

CATERPILLAR

Strongly engaged in climate change public policy, Caterpillar is a member of the U.S. Climate Action Partnership (USCAP), Caterpillar in 2007 called on U.S. policymakers to establish a mandatory emissions reduction program to address climate change—specifically "a federal approach that's well integrated into a harmonized global system to GHG emission-reduction initiatives and avoids local or regional development of separate paths …" Caterpillar's CEO joined other CEOs in USCAP in urging Congress to specify a target zone aimed at reducing emissions 60 to 80 percent by 2050.

"Ultimately," Caterpillar said on its Web site in 2007, "we believe the sustainability of our world and the sustainability of our business are inseparable. The issue of GHG emissions is one significant area in which our core businesses and the challenges of sustainable development merge."

CITIGROUP

Citi announced in 2007 that it will direct $50 billion over the next 10 years to address global climate change though investments, financings and related activities to support the commercialization and growth of alternative energy and clean technology among the clients and markets it serves, as well as within its own businesses and operations.

As an early and substantial force in the world business community for sustainable develop, the Citi organization said it will increase 10-fold, to $10 billion, its commitment to reduce its corporate environmental footprint through its own real estate portfolio, procurement and energy use, as part of its pledge to reduce GHG emissions by 10 percent by 2011.

COCA-COLA COMPANY

Sustainability references and features on Coca-Cola's Web site include these policy and reporting items: "The consensus on climate science is increasingly unequivocal—global climate change is happening and man-made greenhouse gas emissions are a crucial factor. The implications of climate change for our planet are profound and wide-ranging, with expected impacts on biodiversity, water resources, public health, and agriculture.

"Across the Coca-Cola system, we recognize that climate change may have long-term direct and indirect implications for our business and supply chain. As a responsible multinational company, we have a role to play in ensuring we use the best possible mix of energy sources, improve the energy efficiency of our manufacturing processes, and reduce the potential climate impact of the products we sell.

"Regulations to help reduce carbon emissions as a cause of climate change already exist in some of our markets and others are emerging. We believe that, beyond regulatory compliance, business can play a powerful role to help drive climate solutions through innovation and competition.

The company considers its response to climate change as part of its integrated approach to sustainability, stating that "climate protection is a key component of our business strategy, under the Planet element of our Manifesto for Growth."

The Web report states: "The Coca-Cola Company and bottling system, through our Energy & Climate Protection strategy, are committed to finding innovative ways to mitigate the environmental effects of our operations and products. In particular, we are linking climate change to our leadership on water stewardship issues and to our work with supply chain partners on packaging. The three principal activities of the Coca-Cola business system which create greenhouse gases are our manufacturing plants, the distribution fleet, and cold drink equipment. We continually strive to improve the energy efficiency of our plants and fleet. In addition, the area where we have the greatest opportunity to make a difference is cold drink equipment. HFCs (hydroflurocarbons) are one of the six gases which the Kyoto Protocol specifically called to be reduced: We are making good progress toward phasing out HFC use in both refrigeration and insulation as well as to improve the energy efficiency of cold drink equipment by 40–50%."

As an extension of its efforts, Coca-Cola co-founded "Refrigerants, Naturally!" with McDonald's and Unilever along with support from the United Nations Environment Programme (UNEP) and Greenpeace. "This is an example of public-private collaboration on innovative solutions for climate protection," stated the company on its site.

More information on company efforts to cut carbon and to report progress are available through Coca-Cola's environmental reports, responses to the Carbon Disclosure Project, and postings of GHG inventory on the World Economic Forum's Global Greenhouse Gas Register.

DAIMLER

Postings by DaimlerChrysler prior to 2008 described offerings of vehicles in different classes that optimize fuel efficiency, reduce CO2 emissions and improve combustion engines. Its production goal was stated reduction of specific CO_2 emissions in the U.S. production plants by 10 percent between 2002 and 2012, with this comment: "Specific CO_2 emissions were already reduced considerably in the U.S. in 2005. In addition to commissioning numerous feasibility studies on energy supply concepts such as cogeneration, the calculation of CO_2 emissions is currently being standardized and energy benchmarking project is being conducted for paint shops in Germany and the U.S."

Measures on the vehicle side will reduce CO_2 emissions compared with the previous model or the previous baseline version by about 10 to 20 percent, the company said. "The key challenge connected with sustainable mobility is the reduction of fuel consumption and the emission of CO_2 and other pollutants," said the company. "DaimlerChrysler is aware of its responsibility in these areas. In recent years, the Group has already played a major role in helping to make environmentally friendly and sustainable mobility a reality and considerably reducing the fleet's emissions of CO_2 and other pollutants."

The "DaimlerChrysler road map" was presented with these components:
1. Continuous further improvement of combustion engines, with and without a hybrid option.
2. High-quality and alternative fuels.
3. Emission-free driving, with the fuel cell as a long-term goal.

DELL

Dell, in 2007, announced its intention to become the "greenest tech company on the planet." The most far-reaching aspect of its Zero Carbon Initiative was a requirement that its suppliers publicly report their greenhouse gas emissions.

"Suppliers risk having their overall scores reduced during Dell quarterly business reviews for not identifying and publicly reporting GHG emissions," the company stated in 2007. "A supplier's volume of Dell business can be affected by the scores earned on reviews. Dell will work with suppliers on emissions reduction strategies once data is collected.

"We are also taking steps to minimize the environmental impacts of our manufacturing operations. In September 2007, we became the first major computer company to go carbon neutral. Our recycling and reuse rates in our global manufacturing facilities have increased dramatically over the years, reducing the percentage of materials that are land-filled. In fiscal year 2007, we decreased our electricity use by 5 percent in our U.S. facilities compared to fiscal year 2006 and we began to develop similar programs on a global basis."

Dow

In its 2007 Internet position, the company stated: "Dow's vision on overall sustainability is reflected in our 2015 Sustainability Goals—a public commitment that we hold ourselves fearlessly accountable in the pursuit of solutions to the climate change, energy and other pressing world challenges. Our promise is that we will measure and report our progress against this for the next 20 years. Providing humanity with a sustainable energy supply while addressing climate change in the most urgent environmental issue our society faces."

Dow said it will reduce its energy intensity 25 percent from 2005 to 2015.

"Over 90 percent of the products made have some level of chemistry in them, so no one has more at stake in the solution—or more of an ability to have an impact on—the overlapping issues of energy supply and climate change than we do," the company said on its Web site. "As a world leader in chemistry, Dow is uniquely positioned to continue to provide innovations that lead to energy alternative, less carbon intensive raw material sources and other solutions not yet imagined. In fact, our science and technology has been contributing solutions to the global climate change and energy challenge since 1990." Several pages on the site provided details and case examples of the company's efforts.

Duke Energy

As a founder and leading advocate in the U.S. Climate Action Partnership in 2007, Duke Energy expressed its sociopolitical position on climate change:

"We favor a U.S. policy on climate change (that) could be achieved through a carbon tax, through a 'cap and trade' approach, or potentially through other mechanisms. The important thing is that we get to work now. Duke Energy believes that voluntary programs are not enough. Congress needs to establish a national, economy-wide greenhouse gas mandatory program as soon as possible. A sustainable path to reducing U.S. greenhouse gas emissions should become part of a worldwide response to this global issue. We have a responsibility to our customers, our investors and our communities to play a lead role in shaping a national policy that addresses this challenge responsibly and fairly. We must be good stewards of the environment. We must do our part to meet the nation's growing energy needs and to keep our energy prices affordable. We need predictability to make sound plans for electric generation and natural gas infrastructure.

"We are concerned about patchwork policies focused on a single industrial sector or particular region of the country. We are concerned about approaches that could have grave and unintended impacts on the U.S. economy or that could result in rapid or extreme rate increases for electricity and natural gas customers.

"We favor a U.S. policy on climate change that:
- Is economy-wide in its reach, rather than targeting a single industry for emissions reductions;
- Is national in scope, yet considers varying impacts across regions and economic sectors;
- Is market-based, with price signals leading to technological innovation and investment, energy efficiency and conservation;
- Begins to reduce greenhouse gas emissions now, and does so gradually over time;
- Is simple to administer and provides price certainty.

Duke stated that "such a policy could be achieved through a 'cap and trade' approach" which the company, along with others in USCAP advocated in 2007 congressional sessions; and concluded:

"The important thing is that we get to work now. Duke Energy believes that voluntary programs are not enough. Congress needs to establish a national, economy-wide greenhouse gas mandatory program as soon as possible. A sustainable path to reducing U.S. greenhouse gas emissions should become part of a worldwide response to this global issue."

DuPont

Another leader in the U.S. Climate Action Partnership, DuPont advocated climate change policy action in 2007. On its Web site, a robust "Sustainability" section offered details on all the company's economic, environmental and social performance programs and data. Information is extracted from publicly available reports issued by the company to respond to the Global Reporting Initiative surveys. The company's vision ("to be the world's most dynamic science company, creating sustainable solutions essential to a better, safer, healthier life for people everywhere"), environmental stewardship values, sustainability awards and other aspects of its comprehensive "sustainability commitment" are available on the site.

A nine-page speech on DuPont's sustainable-growth platform, delivered by Chad Holliday, chairman and CEO, in October 2006, included these excerpts:

"In 1994, we set a goal to reduce greenhouse gas emissions by 40 percent by the year 2000. We achieved that goal on schedule. Then we challenged ourselves to reduce our greenhouse emission by 65 percent by 2010. We made that goal, as well. In fact, we achieved a 72-percent reduction by 2004, six years ahead of schedule, and avoided costs of over $3 billion by holding our energy use 6 percent below 1990 levels.

"Now we've created a new set of 2015 Sustainability Goals that renew and expand our commitment to sustainability ... By 2015, DuPont will grow our annual revenues by at least $2 billion from products that create energy efficiency and/or significant greenhouse gas emissions reductions for our customers ... By 2015, we will further reduce our greenhouse gas emissions at least 15 percent from a base year of 2004 ...

"For DuPont, 2015 begins today. Sustainable growth is about products and services we are working on right now. Our 2015 Sustainability Goals are our investment in the future of our business ... our customers ... and families around the world. They are about the future of our planet—the one we live on today and the better, safer and healthier planet we aspire to leave for tomorrow."

ExxonMobil

ExxonMobil's energy and climate change section on its Web site provided company information, reports and speeches by top executives on the company's commitments, progress and policy positions.

Remarks by Rex W. Tillerson, Chairman and CEO, Exxon Mobil Corporation, delivered at the Royal Institute of International Affairs, London, on June 21, 2007, are excerpted here: "Climate remains today an extraordinarily complex area of scientific study. The risks to society and ecosystems from increases in CO_2 emissions could prove to be significant—so despite the areas of uncertainty that do exist, it is prudent to develop and implement strategies that address the risks, keeping mind the central importance of energy to the economies of the world. ExxonMobil is taking action to mitigate greenhouse gas emissions today and to support the development of advanced energy technologies with the potential to significantly reduce future emissions. These include: Working with manufacturers of automobiles and commercial industrial engines…Supporting the Global Climate and Energy Project at Stanford University…Mitigating greenhouse gas emissions through efficiency and best practices … Partnering with the U.S. EPA and Department of Energy … Partnering with the European Commission to study carbon capture and storage."

Mr. Tillerson said, "Companies … have a responsibility to take action to combat the rise of greenhouse gas emissions. ExxonMobil is doing so in several substantive ways, consistent with our strategy of increasing efficiency our own energy efficiency in the short-term… advancing current proven emission-reducing technologies in the medium-term … and developing breakthrough, game-changing technologies for the long term. The steps we have taken since 1999 to improve energy efficiency at our own facilities resulted in the avoidance of 12 million tons of greenhouse gas emissions last year alone— the equivalent of taking about two million U.S. cars off the road. "We are partnering with automobile and commercial engine manufacturers on R&D programs that could yield fuel economy improvements in internal combustion engines of up to 30 percent, with lower corresponding emissions. I am pleased to report progress ExxonMobil is making, working with partners in industry and the research community, to develop an innovative fuel system that will generate hydrogen onboard a vehicle as needed. By using liquid hydrocarbon fuels to produce hydrogen, this system is expected to be significantly more fuel efficient than today's internal combustion engines, and also promises reduced emissions without the need for a dedicated hydrogen distribution infrastructure...."

Mr. Tillerson said the company is collaborating with government and science organizations to address climate change issues, citing work with the European Commission to assess the viability of geological carbon storage, based on ExxonMobil's experience in the North Sea Sleipner gas field "where we have sequestered one million tons of CO_2 each year since 1998."

This initiative, said the CEO, will help advance carbon capture and storage technologies which hold promise in reducing emissions in coming decades. "For almost two decades we have funded programs such as the Joint Program on Science and Policy of Global Change at the Massachusetts Institute of Technology," said Mr. Tillerson. "This program brings together a wide variety of scientific, economic, technology and policy experts and integrates the input of a broad spectrum of research to develop comprehensive analyses of climate science and policy. It has made, and continues to make, a vital contribution to our understanding of climate change and the implications for policymaking in this area."

ExxonMobil was a founding sponsor of the Global Climate and Energy Project, based at Stanford University, on research or technologies to meet energy demand with dramatically lower greenhouse gas emissions. Study areas at GCEP include solar power, hydrogen, biofuels, energy storage, carbon capture and storage, and advanced transportation.

As to congressional policy on climate change, Mr. Tillerson said: "... An upstream cap-and-trade system—that is, a system placing a limit on carbon at the point where the fuel enters the commercial world rather than at the point of emission—offers potential advantages in terms of efficiency and simplicity. It reduces the number of regulated entities and provides a cost of carbon to the entire economy. Similarly, a carbon tax could enable the cost of carbon to be spread across the economy as a whole in a uniform and predictable way. Of course, all these policy options carry significant challenges as well as potential benefits, and the devil is very much in the details."

Entergy

In its 2007 Web postings, Entergy was able to put into context for its stakeholders the company's recovery from the devastating impact of hurricanes Katrina and Rita in 2005, providing a lively, comprehensive and personalized sustainability report, accessible as a PDF file.

On the site, J. Wayne Leonard, chairman and CEO, reaffirmed the company's sustainability strength in dealing with environmental, social and economic challenges, to deliver total shareholder return of 38 percent, compared to the 20 percent returned by the Philadelphia Utility Index. In 2006, for the fifth year, Entergy was named to the Dow Jones Sustainability Index- World, which tracks companies that lead their field, as the only company in the U.S. electricity sector. Entergy ranked best in class for social responsibility, corporate governance, climate strategy and other sustainability characteristics. "All of this is evidence," said Mr. Leonard, "that the principles of sustainable growth work—in good times and bad."

A *Forbes* magazine study deemed Entergy as one of the nation's most trustworthy companies "for fair dealing with all stakeholders."

Communication with stakeholders is stressed in Entergy's environmental commitment, presented on the site as corporate efforts to:

- Meet but preferably exceed environmental legal requirements, conforming to the spirit as well as the letter of the law.
- Understand, minimize, and responsibly manage the environmental impacts and risks of our operations, setting goals that reflect continuous improvement.
- Be a good steward of the land that we own and the wildlife and natural resources that are in our care.
- Communicate our commitment to the Policy internally and provide the resources, training, and incentives to carry it out.
- Track and publicly report our environmental performance using best practice reporting guidelines.
- Inform employees, customers, shareholders and the public on matters important to the environment.
- Maintain a constructive dialogue with government agencies and public officials, communities, environmental groups, and other external organizations on environmental issues.

FedEx

An environmental policy statement on the Web site states: "FedEx Corporation and its subsidiaries recognize that effective environmental management is one of its most important corporate priorities...As a company dependent upon transportation, it's particularly important for us to stay at the cutting edge of environmental practices. That's why we are constantly working to create and implement groundbreaking technologies that change the way business is conducted. We are continually and rapidly transitioning to greener practices in all of our operations.."

Some specifics of the commitment: Replacing the FedEx Express Letter with the more environmentally friendly FedEx Envelope, which reduced net greenhouse gases from its production by 12 percent annually; steps to reduce its ecological footprint by using recyclable and non-toxic materials; eco-friendly delivery vehicles such as hybrid electric and delivery tricycles; a fleet of 100 hybrid vehicles; working with industry experts to develop and test hydraulic hybrid technology for the future.

FedEx Corp. is a participant in the Global Environment Management Initiative (GEMI), a non-profit organization of leading companies dedicated to fostering environmental and corporate citizenship worldwide.

Among a large number of environmental awards over several years FedEx lists government recognitions such as the Green Power Leadership Award from the U. S. Environmental Protection Agency in 2001, 2003 and 2005 for Green Power On-Site Generation. ("In 2002, we also earned the EPA's Green Power Partner of the Year"); and the Dubai Cargo Village Environmental Awareness Award in June 2004 for the third year running.

Collaborations with NGOs and others mentioned that "FedEx Express, along with Environmental Defense and Eaton Corporation, received the prestigious Harvard University's Kennedy School of Government biennial 2005 Roy Family Award for Environmental Partnership for their organizations' joint creation of a hybrid delivery truck that reduces particulate emissions by 96 percent and increases fuel efficiency by 50 percent" and "FedEx Express and Environmental Defense awarded the 2005 Blue Sky Award by WestStart-CALSTART for ... nearly single-handed placement of commercial hybrid trucks on the map for corporate America. The FedEx and Environmental Defense effort led the commercial truck market's interest and efforts into the hybrid market. The committee concluded that 'other fleet operators were influenced by the nation's most respected and sophisticated truck operators show strong interest in hybrids ... which in turn motivated those companies to embrace the technologies themselves.'"

FLORIDA POWER AND LIGHT

FPL Group, among the nation's top generating utilities, positions itself on its Web site as "one of the cleanest...producing some of the lowest levels of nitrogen oxides and carbon dioxide per megawatt-hour of electricity." As part of the EPA's Climate Leader program, FPL Group committed to achieve an 18-percent reduction in emissions rates of greenhouse gases by 2008, compared to a 2001 baseline.

FORD MOTOR COMPANY

Ford Motor Company is pursuing a three-pronged climate change/sustainability strategy:
   Continuously reduce the greenhouse gas emissions and energy use of company operations.
   Enhance the flexibility and capability to market lower-greenhouse-gas-emissions products that will attract customers.
   Work with industry partners, oil companies and policy makers to establish an effective and more certain market, policy and technological framework for reducing road transport greenhouse gas emissions.

"Climate change and energy security affect our operations, our customers, our investor and our communities," said Niel Golightly, director, Sustainable Business Strategies. "Some see these challenges as yet another burden on an already stressed industry; but we see potential for innovative business opportunities in solving them."

FORTUNE BRANDS

Company commitment, expressed on its Web site, is embodied a set of principles for its operating companies, concluding with a public policy commitment:

- All operations will be conducted in strict compliance with

applicable environmental laws and regulations.
- Our companies will strive to minimize processes that affect the environment and to reduce the generation of wastes and pollutants.
- Control methods, procedures and processes that are technically sound and economically feasible will be employed to reduce environmental impacts.
- Process innovation will be encouraged to reduce the use of nonrenewable natural resources, the generation of waste and the discharge of substances to the environment.
- Recovery, reuse and recycling options for materials will be considered after efforts to prevent or reduce usage at the source are promoted.
- Employee participation in pollution prevention programs will be encouraged.
- Procurement strategies will be implemented to encourage the use of environmentally friendly products, including the use of recycled, recyclable or reusable products.
- Employees will be educated regarding environmental issues.
- Effective environmental management systems for maintaining compliance and pursuing best practices will be periodically reviewed and enhanced.
- We will communicate with government, industry and the public on environmental issues affecting our businesses and operations as appropriate.
- We will constructively assist all levels of government in the development of equitable and effective environmental laws and regulations, where applicable.

General Electric

GE participates actively in corporate sustainability, both in its business mission and programs inside the company, and in the sociopolitical process that shapes business conditions. CEO Jeff Immelt was among the founders of the U.S. Climate Action Partnership that took the lead in 2007, with green group collaboration, in asking Congress for clear, achievable policy directions. Immelt earlier (May 2005) had launched his company's comprehensive business plan, aimed at environmental and economic goals, known as Ecomagination, billed as GE's commitment to imagine and build innovative technologies that help customers address their environmental and financial needs and help GE grow.

GE's climate change/carbon reduction pledges under its Ecomagination program are grouped under a plan called "1-30-30". Each "30" in the 1-30-30 plan has a percentage meaning: GE commits to reduce the intensity of its GHG emissions 30 percent by 2008, and to improve energy efficiency 30 percent by the end of 2012. The "1" reflects the percentage by which GE will reduce its absolute GHG emissions worldwide by 2012—a big goal given that GHG emissions would otherwise have grown substantially based upon current business growth projections.

GE's approach to a management sustainability strategy process:

Form an internal cross-business, cross-functional team to identify and drive implementation of best practices

Set goals for GE businesses' GHG reductions

Involve top management in business plans to achieve reduction targets

Conduct company-wide communication to engage all employees at all locations worldwide

In its second year, reported on the Web in early 2007, Ecomagination had a portfolio of more than 45 energy efficient and environmentally advantageous products and services. GE announced that it would invest more than $1 billion on cleaner technology research and development (R&D) for 2007, drawing closer to its pledge to invest $1.5 billion annually on Ecomagination R&D by 2010. One of the commitments originally made in 2005, R&D investment, had reached a total of more than $2.5 billion since the program's inception. "GE's ability to reach our ecomagination goals requires a deep commitment to investing in innovative, leading-edge research, while expanding our partnerships with customers, universities and labs," said Lorraine Bolsinger, vice president. "Ecomagination consistently delivers for our customers, and crossing the $1 billion mark for R&D investment will be further evidence of growing momentum."

General Mills

General Mills, a member of The Business Roundtable, the public policy advocacy organization of CEOs from leading U.S. corporations, participates in the Roundtable's Climate RESOLVE initiative which encourages companies to set goals and achieve greenhouse gas reductions. General Mills goal: Reduce GHG generation rate (measured as CO_2 equivalents) by 15 percent over five years using fiscal year 2005 as a baseline.

General Motors

From an announcement in 2007 as the company joined U.S. Climate Action Partnership:

"Today, General Motors announced its goal to reduce CO_2 emissions from its North American manufacturing facilities by 40 percent by 2010, based on 2000 levels. GM is setting this target as part of its voluntary partnership in the U.S. EPA's Climate Leaders program.

When it first joined the program in 2002 as a founding member, GM established a goal to reduce U.S. facility CO_2 emissions by 10 percent by 2005, based on 2000 levels. The company surpassed this target in 2003 with reductions of 11.7 percent. To date, since 2000, GM has reduced U.S. facility emissions by 25 percent, or three million metric tons, which would equal the annual emissions for the power consumed by 288,000 U.S. households.

From an editorial advertising message placed by Chevy on the New York Times op-ed page on January 25, 2008, with reference to the latest General Motors carbon-topic site, www.nytimes.com/chevy:

"At General Motors, we take our responsibility for the environment seriously. If that sounds disingenuous coming from the world's largest producer of cars and trucks, well, maybe it's time to talk—in depth and frankly. Because GM's environmental footprint is indeed large. So is our commitment to change. There is no greater proof of this than the ongoing development of gas friendly to gas-free fuel solutions by our Chevrolet brand. A commitment that has resulted in the Tahoe Hybrid, America's first full-size hybrid SUV, being named the Green Car of the Year by the *Green Car Journal*!

"We expect skepticism. So if you have questions about what Chevy is doing right now with fuel technologies, we want to answer them, right here in this newspaper's Friday Op/Ed section. Simply submit your questions and comments at nytimes.com/chevy. These are important issues. And we are making serious progress. We hope to prove that to you. But we haven't got all the answers. We value your input. And we promise to listen."

The ad was signed by Beth Lowery, GM Vice President, Environment, Energy and Safety Policy.

Hewlett-Packard

HP and World Wildlife Fund-U.S. collaborate in a joint initiative to reduce HP's greenhouse gas emissions from its operating facilities worldwide, educate and to inspire others to adopt best practices (and use HP technology) in conservation efforts around the world. Under the venture, HP pledges to reduce carbon dioxide emissions from HP-owned and HP-leased facilities worldwide to 15 percent below their 2006 levels by 2010. The company will report and verify carbon dioxide emissions from its facilities, based on the Greenhouse Gas Protocol and the World Economic Forum's Global Greenhouse Gas Registry.

HP's most significant source of greenhouse gas (GHG) emissions from operations is energy consumption, primarily electricity. Energy use accounts for 97 percent of GHG emissions. States the HP Web site: "Our environmental operations strategy focuses on consolidating our operations into core sites, with our HP Workplace Transformation initiative enabling us to improve space utilization and install more energy efficient equipment, in a phased approach over time. Our goal is to reduce energy consumption and the resulting carbon dioxide emissions from HP owned and HP leased facilities worldwide to 15% below 2006 levels, by 2010. We are expanding our use of renewable energy, by purchasing green electricity and installing onsite generation at selected locations."

The company presents its full story, dealing with a GHG emissions increase in this manner: "Although our total energy use in operations decreased by 1 percent from 2005 to 2006, our 2006 emissions of greenhouse gases increased 3 percent or by 47,200 tons of CO_2 equivalent. This is due to changes in the global energy mix for electricity generation. In April 2006, the World Business Council for Sustainable Development (WBCSD) and the World Resources Institute (WRI) updated its GHG Protocol emission factors for electricity generation, to reflect these changes. Using the old factors, HP's GHG emissions would have decreased 5 percent." And HP reported in 2007: "Emissions per unit of floor space increased 5 percent due to the changes in electricity conversion factors and a 2 percent reduction in floor space. Emissions per unit revenue, a measure of overall efficiency, decreased by 3 percent."

HOME DEPOT

Environmental principles are described in this statement from its Web site:
- We are committed to improving the environment by selling products that are manufactured, packaged and labeled in a responsible manner, that take the environment into consideration and that provide greater value to our customers.
- We will support efforts to provide accurate, informative product labeling of environmental marketing claims.
- We will strive to eliminate unnecessary packaging.
- We will recycle and encourage the use of materials and products with recycled content.
- We will conserve natural resources by using energy and water wisely and seek further opportunities to improve the resource efficiency of our stores.
- We will comply with environmental laws and will maintain programs and procedures to ensure compliance.
- We are committed to minimizing the environmental health and safety risk for our associates and our customers.
- We will train our employees to enhance understanding of environmental issues and policies and to promote excellence in job performance and all environmental matters.
- We will encourage our customers to become environmentally conscious shoppers."

HONEYWELL INTERNATIONAL

Through its participation in the Business Roundtable's Climate Resolve Program Honeywell is quantifying greenhouse gas emissions, evaluating options for greenhouse gas reductions and setting objectives with respect to those options. HP stated on its 2007 Web site: "We will continue to develop products and technologies that improve efficiency and lower greenhouse gas emissions."

As an example of the company's many moves that link to carbon reduction and climate change, in which products and technology come into plan, HP stated: "To meet the growing demand for high-quality, low emission biofuels, Eni S.p.A., a leading European refiner, is building a production unit in Livorno, Italy using the UOP/Eni Ecofining™ technology developed by the two companies. The facility, which is slated to come on line in 2009, will process 6,500 barrels per day of vegetable oils to produce 'green' diesel fuel. The UOP Ecofining process is a catalytic hydro processing technology that converts vegetable oils and wastes into green diesel fuel ... Eni plans to use the high-cetane green diesel it will produce as a blending stock to enhance the quality of its existing diesel supply."

IBM

The company's first formal environmental and energy corporate policies date back to 1971 and 1974 respectively, and programs supporting them have been embedded within IBM's corporate wide environmental programs and global environmental management system since that time. IBM has comprehensive and multifaceted programs focused on energy efficiency and climate protection.

IBM states: "Climate change is one of the most critical global environmental challenges facing the planet. At IBM, we believe the most proactive approach the company can take to address the complex issue of climate change is to apply our technological and engineering leadership to reduce emissions associated with our own operations and to create innovative products and solutions that are increasingly energy efficient, extending this environmental benefit to our clients. And this is something IBM has been doing and has sustained for many years."

Areas cited by IBM related to climate change:
- Reducing greenhouse gas emissions associated with the company's operations by:
- Conserving energy
- Reducing perfluorocompound (PFC) emissions
- Using renewable energy

- Supporting alternate employee commute options
- Increasing the efficiency of the company's logistics
- Developing energy efficient products and providing diverse solutions for energy efficient data centers
- Collaborating with its clients and others on innovations that help protect the world's climate

From 1990 to 2006, IBM avoided nearly 3 million metric tons of carbon dioxide (CO_2) emissions, equal to 44% of the company's 1990 global CO_2 emissions, and reporting saving more than $290 million through its annual energy conservation actions. IBM set a new goal in 2006 to reduce CO_2 emissions associated with its energy use 12 percent between 2005 and 2012 through energy conservation, the use of renewable energy and/or the funding an equivalent CO_2 emissions reduction by the procurement of Renewable Energy Certificates (RECs) or comparable instruments.

INTEL CORPORATION

Intel and EPA announced on January 28, 2008 that Intel had become the single largest corporate purchaser of renewable energy certificates in the U.S.

The company's Web site states: "At Intel, we've made energy conservation and support for renewable energy a top priority, and we're proud to be part of the EPA's Green Power Partner program. Our purchase of renewable energy certificates (RECs) amounts to 1.3 billion kilowatt hours of energy a year. The EPA estimates this has the equivalent effect of eliminating CO_2 emissions for more than 185,000 automobiles or the electricity needed to power more than 130,000 average American homes annually."

Intel Corporation and Google joined with Dell, EDS, the Environmental Protection Agency, HP, IBM, Lenovo, Microsoft, Pacific Gas and Electric, World Wildlife Fund, and more than a dozen additional organizations announcing their intent to form the Climate Savers Computing Initiative. The goal of the broad-based environmental effort is to save energy and reduce greenhouse gas emissions by setting aggressive new targets for energy-efficient computers and components, and promoting the adoption of energy-efficient computers and power management tools worldwide.

Intel is in EPA's Climate Leaders program, an industry-government partnership under which companies develop goals and strategies aimed at reducing their overall climate change impact. "We believe global climate change is a significant issue and we have been taking steps to reduce our impact on climate change for many years," said Todd Brady, Intel's Corporate Environmental Manager. "Participating in EPA's Climate Leaders program will help raise awareness for this issue, and allow us to share our learning with other committed companies."

JOHN DEERE

Deere & Company puts its current sustainability achievements into a 30-year context. The company has had an energy management program since 1973 requiring operations to implement energy-conservation initiatives and to track energy use. In 2003 Deere added a worldwide greenhouse gas inventory program.

As part of the voluntary U.S. Environmental Protection Agency's Climate Leaders program, "representatives from Deere's 15 largest global manufacturing facilities are working together to reduce emissions of greenhouse gases such as carbon dioxide," states the company Web site.

Its 32-page Global Citizenship Report (2007) is available on the site. Some excerpts:

At Deere's new Brazil factory, for every tree taken out, four more were planted. "Minimal-energy-use principles were followed to select energy-efficient motors, lighting design and the facility's energy management systems. The use of a retention pond controls storm water runoff. The plant lacks underground oil or chemical storage tanks or floor drains and has installed wells for groundwater monitoring." John Deere's facility in Pune, India is described as "a zero-wastewater discharge facility."

Available on the site is CEO Robert W. Lane's 2006 speech at a renewable energy conference in St. Louis, showing Deere's high-level commitment to growing sustainable businesses, and urging public policy that encourages initiatives such as renewable energy.

"Sustainability," said Deere's leader on that occasion, "requires that the business provide a high-quality product or service that fully meets the customer's needs; a workplace environment that attracts the best and brightest employees and enables them to realize their potential; and, succeeding with the customers and employees, in this way is a necessary condition for ensuring that shareholders receive attractive returns on their investment. In short, sustainability requires laying building blocks in the beginning that will serve these vital constituencies well."

On public policy encouragement, Lane said: "Public policies can provide useful "pump priming" for new, evolving industries, creating the infrastructure needed for markets to develop and businesses to grow. Tax and regulatory incentives can provide needed planning stability while business conditions evolve, facilitating the growth of new business ventures. Over time, as markets and businesses mature, the need for these incentives will recede and a transition away from these policies is appropriate."

JOHNSON & JOHNSON

The company's Web site states: "As indicated in our Next Generation Goals, adopted in 2000, it is the responsibility of each company/business unit to meet our greenhouse gas reduction goal of 4 percent by 2005 and a 7 percent reduction by 2010, in absolute terms with 1990 as a base year. The pathways for a climate-friendly energy policy include five elements:
- Energy efficiency improvements in all of our operations
- Cogeneration: on-site generation of electricity and recovery of the waste heat for overall efficiencies of 80+ percent
- On-site renewable energy that produces no CO_2 emissions
- Renewable electricity purchases
- Carbon trading and sequestration"

Johnson & Johnson has had environmental goals for more than 15 years, setting new long-term goals every five years. We have reduced our environmental impacts significantly during this time. In 2006, we embarked on our newest 5-year goal period. The Healthy Planet 2010 goals (2006 to 2010) were developed after extensive stakeholder engagement with government representatives, nongovernmental organizations, other companies, academic thought leaders and Johnson & Johnson Family of Companies employees at all levels.

J&J's 2006 performance against these goals (reported in the company's 2006 Sustainability Report, available online) is reported as follows on the Web:

Energy Use – Carbon Dioxide Reduction Goal: Absolute reduction in CO_2 emissions of 7% from 1990–2010 Actual: 16.8% decrease from 1990–2006. Goal: Reduce fleet total CO_2 emissions per mile driven by 30%. Actual: Minimum fuel efficiency requirements for our fleet were established by vehicle category. In addition, the U.S. fleet had 554 hybrid vehicles at year-end 2006. ("As yet," said the J&J report, "we have not seen an overall improvement in emissions per distance driven.")

Monsanto

Monsanto, long a leader in overcoming sociopolitical challenges facing the chemical industry, has moved steadily toward sustainable agricultural practices and products. The company's Web site addresses climate change as "among the most pervasive threats to the web of life", and provides this educational comment: "Although there is still some debate about the causes and ultimate effects of climate change, humans have the power to address at least some of its sources. The environmental benefits of no-till (farming) are significant. Farmers who use no-till leave the remains of the previous crop as mulch on the field. This mulch protects the soil from wind and rain, and it recycles nutrients into the soil to help the next season's crops grow. Moreover, carbon from the retained plant mater is recycled into the soil (and sequestered)."

Monsanto announced in 2007 that it had joined the Chicago Climate Exchange (CCX), North America's only voluntary, legally-binding greenhouse gas emissions reduction, registry and trading program. As part of its agreement, the company will by 2010 reduce its own direct carbon emissions from major U.S. operations by 6 percent below its 2000 levels or purchase carbon emission offsets as specified in the CCX contract. The company said it would also work with farmer groups to discuss reducing carbon dioxide in the air by practicing no-till agriculture, which involves minimal plowing of farmland. This practice sequesters carbon in the soil rather than releasing it into the air in the form of carbon dioxide.

Navistar

Navistar believes that achievement of the vital public policy goals of energy efficiency and greenhouse gas emissions reduction will require strong collaboration of public and private sector interests, and the company expects to be proactive in that process.

"Navistar's position on carbon constraint is that public policy mandates must recognize the significant fuel-efficient, lower-carbon contribution of clean diesel technology. We believe that government at all levels should encourage greater use of diesel transportation, advanced diesel technology and emerging solutions such as diesel hybrid engines as an important part of the answer to global climate change. Our company is committed to sustainable greenhouse gas reduction in our plants and in our products. Navistar's facilities continuously improve energy efficiency through enhancements in lighting, insulation, technology, equipment and operating practices. Our comprehensive emission inventory system enables us to track our greenhouse gas reduction progress. (Information on this is available in our annual sustainability report, online at Navistar.com, and in our reports to the Carbon Disclosure Project.)

"In our products—diesel engines, trucks, school buses and emission-reduction retrofits—Navistar leads by example. Our Green Diesel Technology® engines were the first to achieve current federal standards for diesel trucks and school buses, cutting black carbon levels to near zero. Our diesel hybrid vehicles emit 40 percent less carbon dioxide than conventional diesel engines, and our aerodynamic vehicle designs will further reduce emissions of this greenhouse gas. Our vehicles are able to use bio-diesel fuel to ease petroleum reliance.

"Navistar, with other business interests and through our own customer education programs, will continue to raise awareness and encourage responsible action on diesel-related environmental matters. As a member of concerned organizations such as the Business Roundtable (participating member of the Climate Resolve program) and the National Association of Manufacturers, Navistar will work collaboratively with public policy makers and all our stakeholders to achieve rational and sustainable solutions related to climate change."

NESTLÉ

The company's Web site provides information on operations, as part of the Nestlé Environmental Management System, which is aligned with the international standard ISO 14001, where Nestlé systematically reviews factory efforts to minimize air emissions. An entry in 2006-

2007 stated: "The air acidification potential from Nestlé manufacturing operations, measured in sulfur oxide equivalents, reduced significantly by 14.5 percent between 2004 and 2005. Since 1998, it has reduced by more than 62 percent per ton of product.

"The CO_2 emissions from our manufacturing operations were also reduced by more than 12 percent since the start of negotiations leading to the 'Kyoto Protocol' in 1997. During this period, our production volumes increased by more than 55 percent. Between 2001 and 2004, Nestlé reduced GHG emissions per ton of product by 17% in Latin America."

News Corporation

Chief executive Robert Murdoch launched a company-wide plan to address climate change that includes not only a pledge to reduce the company's emissions … but also a vow to weave climate messaging into the content and programming of News Corp.'s many holdings. Murdoch said that News Corp.'s hundreds of millions of viewers and readers represent the most fertile ground for change.

"Imagine," said Murdoch, "if we succeed in inspiring our audiences to reduce their own impacts on climate change by just 1 percent. That would be like turning the state of California off for almost two months.

"We're not an industrial company or an airline, but we do use energy in our activities: publish newspapers, produce films and television programs, and operate 24-hour newsrooms and others. We want to make energy efficiency part of our everyday operations and switch to renewable sources of energy wherever economically feasible.

"We intend to reduce our use of energy and find energy from renewable sources—enough to decrease our carbon footprint in 2012 by 10% compared with 2006.

"We plan to become carbon neutral by 2010 by offsetting emissions we can't avoid."

Pacific Gas and Electric

"We are committed to leading by example when it comes to climate change. That means more than just minimizing the greenhouse gas emissions from our operations. It also means maximizing the opportunity to we have to lead on efforts to establish responsible policies and programs to address global climate change," PG&E states on its Web site.

PG&E received third-party certification for its entity-wide CO_2 emissions for 2002 and 2003 using the California Climate Action Registry's rigorous reporting standards and protocols. These standards and protocols are informed by the World Resources Institute/World Business Council on Sustainable Development's Greenhouse Gas Protocol. We support mandatory, national emissions-reduction efforts that are market-based, encourage efficiency, and promote new technologies.

On June 28, 2007, PG&E launched ClimateSmart, a voluntary program that will allow its customers to take action to reduce greenhouse gas emissions and make their energy use "climate neutral." Through the program, customers can choose to pay a small amount on their monthly utility bill to fund new environmental projects aimed at removing an equal amount of carbon dioxide from the atmosphere.

And, said PG&E, "we continue to reduce the methane leak rate from our natural gas pipeline operations, avoiding the release of more than 670 tons of methane, or approximately 14,060 tons of CO_2-equivalent, in 2006." At the national level, Peter A. Darbee, Chairman, CEO and President of PG&E Corporation, participated in the U.S. Senate's Committee on Environment and Public Works hearings on climate change in February and June of 2007. In a news conference later that year, Darbee announced the launch of Combat Climate Change (3C), an international business initiative that brings together 46 leading international companies. In advance of the 2007 climate change negotiations in Bali, Indonesia, the group called on governments to work together to develop a global policy framework to combat climate change.

PG&E Corporation is a member of the United States Climate Action Partnership (USCAP)—a group of businesses and leading environmental organizations that called on the federal government to quickly enact strong national legislation to require significant reductions of greenhouse gas emissions. USCAP has issued a landmark set of principles and recommendations to underscore the urgent need for a policy framework on climate change.

PG&E explains its political involvement in climate change: "We were an original member of the Clean Energy Group, a coalition of environmentally progressive utilities advocating for a national, mandatory, market-based approach to curbing greenhouse gas emissions in the power sector. And while we prefer a national approach to addressing climate change, we also recognize the important role that states can play—in terms of acting as a catalyst for federal legislation and bringing forth innovative ideas and approaches that tackle the complexities of the issue. That is why we worked constructively with the California legislature and the Administration to enact AB 32, the Global Warming Solutions Act, as well as SB 1368, which requires that all power sold to utilities in the state under long-term contracts meet a greenhouse-gas emissions performance standard that is equivalent to that of an efficient combined cycle natural gas plant. Implementation of these two pieces of legislation will ensure that California continues on a path of reducing its overall carbon footprint, and challenging its businesses to do the same."

PFIZER

Pfizer in 2007 set a companywide goal to "reduce carbon dioxide emissions by 35 percent per $1 million of sales (from a 2000 baseline) and, by 2010, supply 35 percent of our global energy needs through cleaner sources."

A member of EPA's Climate Leaders program, a voluntary industry-government initiative, Pfizer states:

"Our long-term focus on reducing emissions to the environment has caused us to significantly reduce releases to the environment despite company growth. Since 1994, Pfizer has reduced its emissions of solvents to the atmosphere and wastewater by 70 percent...This was done through improving processes using green chemistry principles and ensuring that all production facilities are equipped with advanced emission controls."

Pfizer listed these "goals and commitments" on its Web site in 2007:

- To meet 35 percent of our electricity needs by 2010 through the use of "cleaner" energy technologies, such as co-generation.
- To reduce CO_2 emissions by 35 percent per million dollars of sales by 2007 from baseline year 2000.
- To reduce the impacts of our nearly 38,000 automobiles worldwide.
- To effectively manage the financial implications and opportunities associated with the energy reductions gleaned from our conservation and clean energy projects.
- To identify the operating risks and business opportunities presented by a changing global climate.

Procter & Gamble

Procter & Gamble has a long record of leading business sustainability concepts, dating back to the early 1990s when George Carpenter and other pioneers in corporate greening worked together to make company intentions and records understandable and transparent.

P&G has recently underscored its belief in the "growing scientific evidence substantiating links between greenhouse gas emissions and global climate change" and says (on its Web site): "We are concerned about the potentially negative consequences, and believe prudent and cost-effective action to reduce the emissions of greenhouse gases to the atmosphere is justified, under the consensus of the U.N. Framework Convention on Climate Change; [however,] P&G is not an energy intensive company... any voice P&G could have in the policy debates surrounding climate change will be minor."

PG&E is a member of the U.S. Business Round Table's Climate RESOLVE program, which was created in response to President Bush's challenge for U.S. businesses to voluntarily reduce overall greenhouse gas emissions intensity by 18 percent by 2012.

Prudential

Prudential web site states that "Our programs for reducing energy consumption have already cut net greenhouse gas emissions at our major U.S. facilities by 33 percent since 1998. We plan to further reduce our emissions. We also have active recycling programs in our major facilities and are committed to using recycled paper products in materials such as our annual report."

Raytheon

In the "stewardship" report posted on its Web site, Raytheon acknowledges the role of human activity in greenhouse gas emissions and climate change, states that the company has had a strong energy conservation program since the 1970s, and that more than 90 percent of the greenhouse gas (GHG) emissions attributable to the company are from energy consumption, primarily from purchasing electricity from third-party power plants.

Highlights of Raytheon's energy initiatives:

Guide to Corporate Climate Change and Sustainability Positions

- Since 1999, Raytheon has been an ENERGY STAR Partner, committed to improving energy conservation performance under this joint program of the U.S. Environmental Protection Agency (EPA) and the U.S. Department of Energy.
- In 2002, Raytheon joined EPA's Climate Leaders program as one of its charter member companies. Climate Leaders is a voluntary industry/government initiative that requires participating companies to develop comprehensive, long-term GHG reduction strategies.
- Raytheon set an aggressive goal to reduce GHG emissions by 33 percent from 2002 to 2009, normalized by revenue. By the end of 2006, we achieved a 24 percent reduction toward this goal. Since 2002, we have eliminated cumulatively 155,000 metric tons of carbon dioxide equivalent GHG emissions.
- Recognizing the growing need to conserve energy, we set a related challenge of reducing electricity consumption by 15 percent from 2005 to 2007, adjusted for business growth. By the end of 2006, we achieved an 8 percent reduction toward this goal, saving $9 million in costs, and 80 million kilowatt hours of electricity—enough energy to provide power to 8,000 homes for a year.
- Energy conservation results are driven by the leadership of our Enterprise Energy team and more than 900 Energy Champion volunteers. Energy Champions have the passion and drive to conserve energy through their own initiatives and by influencing peers in the workplace to do the same.
- We expect our newly designed and constructed Huntsville, Ala., building to be certified as a new LEED building. In 2004, one of our Marlborough, Mass., buildings was certified as a new LEED-green building.
- We are studying the feasibility of wind turbines for our Portsmouth, R.I. location. We also have plans to construct a pilot photovoltaic (solar electricity) system on the roof of our building in Andover, Mass. These two examples of sustainable, renewable energy demonstrate our broad commitment to further reduce GHG emissions and our dependency on electricity generated at power plants.

Rio Tinto

"We believe that the activities of people and companies are causing emissions of greenhouse gas which are contributing to climate change. So we realize that we must play our part in reducing global greenhouse gas emissions," this international mining company stated in 2007. "We also accept the work of the Intergovernmental Panel on Climate Change (IPCC) as a starting point from which to develop our policies. Our climate change position guides our work to reduce greenhouse gases.

"We have just completed one year of our current three-year Climate Change Plan, which takes a stepped, group-wide approach to addressing each theme ... Our Climate Change Leadership Panel is responsible for ensuring that our program remains on course. All of our business units are implementing their own programs in support of the plan and each business has a climate change champion. We also hold regional workshops to encourage the sharing of information and best practice.

"We have been reporting our GHG emissions publicly since 1996. In addition to our annual external verification of health, safety and environmental data, we have participated in the Australian Greenhouse Challenge verification program. In addition, external consultants have, on two occasions, reviewed the methodology we use against the standards of the IPCC and the World Business Council for Sustainable Development.

"Our greenhouse gas emissions are largely dependent on how well we manage our energy use. To that end, we have instituted a comprehensive program of energy audits at our operations to identify a range of energy saving opportunities, and several of these have already been successfully implemented. In 2004, we set five year targets to reduce our greenhouse gas emissions by four per cent per ton of product by 2008 (using a 2003 baseline) and to reduce energy use per ton of product by five per cent per ton of product over the same period.

"We are also a leading international coal producer, and so we are helping to develop technology for carbon capture and storage worldwide. In addition, we are a significant supplier of uranium oxide for the world's nuclear industry, which is a low emitter of greenhouse gases. Along with developing processes for product stewardship, we have formed with other uranium producers a group to look at the implications of the entire fuel cycle."

Shell

"Contributing to sustainable development is not only the right thing to do," the company states on its Web site; "it makes good business sense. Sustainable development helps us to be a more competitive company and create value for our shareholders."

Shell provides an insight for Corporate Greening 2.0 in comparing the green challenges of previous years with those associated with climate change, with the consistent necessity to maintain economic sustainability. "Somehow," Shell states "the emission and negative social impacts from fossil fuels will need to go down, even as the use of coal, oil and natural gas continue to rise. We have done it before—with local air pollution…In 1997, Shell was one of the first energy companies to acknowledge the threat of climate change; to call for action by governments, our industry and energy users; and to take action ourselves. This continues to be reinforced by our senior executives today. Climate change is the …biggest challenge yet. And doing again what we have done for air pollution with greenhouse gas emissions will be a real challenge."

The company addresses action on many fronts, from improvements in energy efficiency and increased use of renewable energy, to large-scale CO_2 capture and storage from fossil fuels and a slowing of deforestation. Here are other excerpts from Shell's extensive sustainability communications, both on its Web site and in public presentations:

"Fossil fuels will continue to provide the majority of the world's growing need for energy for decades to come. They remain the most convenient and affordable source of energy we have and the only one currently available on the enormous scale needed.

"That makes managing CO_2 emissions from coal, oil and natural gas critical to addressing man-made climate change. We invest in technologies and projects to provide the extra energy societies need and demonstrate the integrated CO_2 solutions they expect.

"The challenges surrounding climate change cannot be met by energy producers, industry or consumers alone. Governments must set the framework to encourage investments in low and zero CO_2 emitting energy and promote energy efficiency and responsible energy use. We have stepped up our appeals to governments to lead on this issue and introduce effective, policies, like emissions trading, to combat climate change.

"We are taking action by managing GHG emissions in our worldwide operations. In 2006, facilities we operate emitted 98 million tons of GHGs, about seven million lower than the previous year and more than 20% below 1990 levels ... We set an aggressive voluntary target to reduce GHG emissions from our operations in 2010 to a level 5% below 1990 levels, even while growing our business. By the end of 2002 we had met our first GHG target, reducing emissions by over 10 percent compared to 1990, through efforts such as the elimination of continuous venting.

"Our customers emit six to seven times more CO_2 using our products than we do making them—more than 750 million tons of CO_2 in a typical year. We are encouraging people to use energy efficiently and providing them with electricity and transport options emitting less CO_2."

A statement from Jeroen van der Veer, Shell's chief executive, endorsing government help on market-based policy, was posted in 2007:

"For us, as a company, the scientific debate about climate change is over. The debate now is about what we can do about it. Businesses, like ours, should turn CO_2 management into a business opportunity and lead the search for responsible ways to manage CO_2, use energy more efficiently and provide the extra energy the world needs to grow. But that also requires concerted action by governments to create the long-term, market-based policies needed to make it worthwhile to invest in energy efficiency, CO_2 mitigation and lower carbon fuels. With fossil fuel use and CO_2 levels continuing to grow fast, there is no time to lose."

Siemens

"For us," the company states on its Web site, "sustainable development in environmental protection means careful use of natural resources, which is why we assess possible environmental impacts in the early stages of product and process development...Issues like climate change and the responsible use of critical resources like water are relevant to our production and office sites."

News available on the Web site includes the commitment of Siemens to the Clinton Climate Initiative, which includes a transportation program to provide cities with access to energy-efficient and money-saving transportation products and solutions." "At Siemens we recognize the impact that major cities and the vehicles within them have on climate change, which is why we are so honored to partner with the Clinton

Climate Initiative," said Dennis Sadlowski, President and CEO, Siemens Energy & Automation, Inc. SE&A has implemented thousands of energy-efficient drive solutions around the globe, said the company; by leveraging this experience, the program can reduce fuel consumption and emissions.

The Web site provides several examples of commercial activities related to climate change and sustainability, including conversion of a boiler in a heating power station in Bavaria to natural gas, that lowered emission of carbon dioxide by 45,000 metric tons during the 2005 heating season; Norway's largest wind park; and supplying biomass power plants that produce energy from renewable resources.

"The flexibility of Siemens' ELFA® modular and highly efficient hybrid drive systems allows the combination of nearly all common energy sources and storages. The ELFA system synchronizes the two power systems via its power electronics to optimize the energy flow, resulting in up to 40 percent less energy consumption and CO_2 emission as well as quieter and more comfortable and reliable operation," the company states. Siemens supplies its ELFA systems to bus manufacturers, who install them into vehicles and sell the complete bus to the transportation authorities.

Sony Electronics

Sony has pledged to protect and improve the environment in all area of its operations, while acknowledging that conserving resources, increasing efficiency and reducing pollution also lower the company's operating costs.

The company Web site provides a long-term sustainability perspective, noting that Sony Corporation of America in 1994 strengthened its environmental commitment with a green action plan for its electronics and entertainment businesses. This led to the establishment of Sony Environmental Vision Green Management 2005.

The plan creates a framework, unique to each business, to uphold green policies and procedures in the creation, design, manufacture, packaging, transport, sales and service of products and in all areas of operation.

Toyota

Toyota's sustainability flag is shown in this paragraph on its Web site: Green. That's how we'd like the world to be. As an environmental leader, Toyota does more than meet industry standards—we seek to raise them. With an unwavering commitment to environmental protection, Toyota strives to create clean and efficient products, and to conserve resources before our vehicles even hit the road.

Toyota has set five environmental goals to help mitigate the company's energy and greenhouse gas (GHG) footprint: improve fuel efficiency; promote fuel diversification; develop advanced vehicle technologies; promote advanced transportation solutions; and reduce energy and GHG emissions across company operations.

Providing details, the Toyota site stated in 2007: "At Toyota, we operate under a global Earth charter that promotes environmental responsibility throughout our entire company. We are leading the way in lowering emissions and improving fuel economy in gasoline-powered vehicles. In addition, as part of our dedication to environmental preservation, we have developed strong partnerships with organizations like The National Arbor Day Foundation and The National Environmental Education & Training Foundation.

"The Global Earth Charter, under our guiding principles, was set forth to promote environmental responsibility for every aspect of our company and significantly reduce the impact our vehicles have on the planet. That's why we subscribe to a recurring 5-year Environmental Action Plan that sets Earth-friendly goals. Toyota is happy to report that we've successfully achieved our first Action Plan for U.S. operations, and have now launched our second.

"Clean energy—and ensuring its abundant supply to meet the world's future needs—will be one of the defining challenges for the 21st century. Society's energy demands continue to rise, particularly here in North America. Right now, our energy is sourced from carbon-based fossil fuels such as petroleum, coal and natural gas. In North America, 41% of our energy comes from petroleum-based fuels; approximately two-thirds of that petroleum is used for the transportation sector...Fossil fuels are essentially nonrenewable, becoming harder to extract, and much of it comes from outside North America. For all these reasons, petroleum-based fuel is becoming more expensive, as we have seen as the price of gas has risen at the pump. When we drive a vehicle, it consumes fossil fuels and emits CO_2, a major contributor to climate change. So neither the feedstock of the conventional car—fossil fuels—nor the consequences of its use—climate change—are sustainable models for vehicles in the future. We need to design transportation solutions that overcome our reliance on fossil fuels.

"Energy needs and climate change are complex issues. They will require societal action—the combined efforts by governments, policymakers, corporations and individuals—to address them. There is no single 'silver bullet' solution—a multi-prong strategy is needed. The auto industry can accelerate the availability of fuel alternatives; we can improve the energy efficiency of our business operations; we can offer more fuel-efficient technologies and products. However, the broad commercialization of these ideas will require commitment from groups outside of the auto industry. Energy providers will need to provide energy from renewable sources. Fuel providers will need to make new fuels available and provide the necessary distribution infrastructure. Government will need to establish incentives that spur development and the purchase of new technology and low carbon energy. Consumers will need to push market demand. In short, our success will be dependent on change from all sectors.

"Toyota recognizes the growing need to take action to promote energy diversity and address climate change. We are not waiting for others to act before we take action ourselves. We are conducting a broad North American Greenhouse Gas inventory to understand the current GHG footprint of our operations and products. We are investing time, funding and our experience in collaborative and policymaking efforts to respond to climate change and help to diversify energy sources. We are developing cars and trucks that can travel farther on a single tank of fuel and operate on a variety of clean energy sources. We are becoming more efficient in how we design, build, distribute and sell our products. Our goal is to develop innovative technologies for the future while continuously improving the mainstream technologies of today in a way that meets customer needs and brings us closer to sustainable mobility."

UPS

On two of its sites in April 2008 (http://www.community.ups.com/ and http://sustainability.ups.com/), UPS provides its environmental and sustainability commitments and record. The company speaks to its stakeholders in providing details the status of key green and community performance indicators as of year end 2006. "We've provided our stakeholders with a chart that outlines our 2002 baseline data, 2006 status and 2007 goal for each KPI. We've also updated each section of this site with new information about major initiatives launched in 2006 to maintain or improve our performance," the company states. The chart opens to show details in each category.

UPS states its triple bottom line commitment ("At UPS, we believe our business success depends upon balancing economic, social and environmental objectives") and continues:

"Today, synchronizing commerce is more than a business process or an emerging industry space. It's about operating in unison with employees, communities and governments to foster greater global economic prosperity and encourage individual achievement. It's also about providing optimal service and value to our customers by striving for the highest operational efficiencies and minimizing our impact on the environment. We believe this long-term, sustainable approach to running our business benefits UPS, our employees, customers, share-owners and the communities in which we operate..."

UPS says its company-wide environmental management system includes focuses on environmental inspections by government agencies, procedures regarding underground petroleum storage tanks, management of incidental spills, and monitoring aircraft deicing runoff. Patterned after the ISO 14001 standard, UPS's environmental management system provides guidance to plant engineering staff regarding their responsibilities for regulatory and waste minimization programs. "We have 440 full-time equivalent employees whose responsibilities include carrying out our environmental programs, processes and activities in accordance with regulatory and UPS-specific requirements. While we comply with all applicable government regulations, we also exceed requirements on many initiatives. Our training and auditing programs identify areas for improvement and outline strategies for achieving it. More than 1,000 employees have received training since 1991, and results of annual audits are reported to UPS's senior management."

On GHG emissions: "As new fuel efficient and alternative technologies become widely available and affordable, UPS's long-term goal is to decrease total CO_2 emissions produced by our operations. In the near term, our efforts are focused on reducing emissions per package."

In March 2008, UPS announced the deployment of 167 Compressed Natural Gas (CNG) delivery vehicles in Texas, Georgia and California, to help reduce the company's carbon footprint and dependence on fossil fuels.

UNITED TECHNOLOGIES

The focus of UTC's programs is to reduce the negative environmental impacts of operations, particularly those than may affect global climate change and the ozone layer. UTC was a founding member of the Pew Center on Global Climate Change Business Environmental Leadership Council and has been an EPA Climate Leader since 2003.

UTC's 2007 Web site stated: "In 2006, we joined the Chicago Climate Exchange as a Phase 1 and Phase 2 member. In 2007, we embark on a new four-year program to reduce our greenhouse gas emissions by 12 percent…on an absolute basis compared to 2006. A significant change is that the (metrics) will no longer be normalized for volume and instead will be tracked in absolute terms. Our greenhouse gas reduction target, equivalent to taking more than 50,000 cars off the road, is aggressive, as our performance to date has averaged 2 percent reduction annually."

UTC released its 2007 corporate social responsibility report in February 2008 with these highlights in a news release:

"The company's environmental objective is to cut greenhouse gas emissions three percent annually and water consumption 2.5 percent annually from 2007 to 2010. [In 2007, the company says it cut greenhouse gas emissions five percent and water consumption six percent.]

"The company had also set a goal of investing $100 million from 2007 to 2010 in energy conservation projects, including co-generation systems." (In 2007, UTC said it identified more than 600 projects valued at $80 million and funded $31 million, half of which supports co-generation systems.)

In 2007, more than 1,000 suppliers representing approximately 40 percent of UTC's product spend completed self-assessments against baseline EH&S expectations. Eighty-one percent have met the criteria. [UTC says that it aims to have 100 percent of corrective actions closed by the end of 2010.]

WAL-MART

Wal-Mart's aggressive position was staked out when its CEO appeared on the cover of Fortune magazine in 2006 with this opening paragraph: *Lee Scott is no tree-hugger. But Wal-Mart's CEO says he wants to turn the world's largest retailer into the greenest. The company is so big, so powerful, it could force an army of suppliers to clean up their acts too. Is he serious?* [122]

Green strategies were structured and are being implemented toward serious sustainability; and Wal-Mart had this to say about climate change on its 2007 Web site:

"Our planet's future depends on our redesigning the current energy system, which relies on fossil fuel that emit tremendous amounts of carbon and greenhouse gases into the air, which many scientists say is altering our climate. Climate change is an urgent threat, not only to our business, but also to our customers, communities, and the life-support systems that sustain our world.

"As significant consumers of energy worldwide, we are committed to doing our part. Through deep investments and efficiency innovations in our stores and trucking fleet, we plan to reduce our overall greenhouse gas emissions by 20 percent over the next eight years. We will also design a store that will use 30 percent less energy and produce 30 percent fewer greenhouse gas emissions than our 2005 design within the next three years."

The Clinton Climate Initiative in 2007 announced a partnership with Wal-Mart to help bring environmentally-friendly technologies to cities across the United States and around the world, to explore ways to use purchasing resources to lower prices on sustainable technologies such as energy-efficient building materials and systems, energy-efficient lighting and clean energy products.[123]

Also in 2007, Wal-Mart President and CEO Scott unveiled "Sustainability 360," a company-wide emphasis on sustainability extending beyond Wal-Mart's direct environmental footprint to engage

---

[122] Article by Marc Gunther, *Fortune Magazine*, July 2006.

[123] Former President Bill Clinton made the announcement at the United States Conference of Mayors Climate Protection Summit held in Seattle. In addition to the partnership with Wal-Mart, Clinton announced that the Clinton Climate Initiative would extend its programs and purchasing consortium to all 1,100 cities represented by the Conference of Mayors. The consortium includes the C40 Climate Leadership Group, representing 40 of the world's largest cities.

associates, suppliers, communities and customers. The announcement was made during Scott's keynote lecture at the Prince of Wales' Business and the Environment Program in London.

"Sustainability 360 takes in our entire company -- our customer base, our supplier base, our associates, the products on our shelves, the communities we serve," said Scott. "And we believe every business can look at sustainability in this way. In fact, in light of current environmental trends, we believe they will, and soon."

Appendix

Climate Change Advice to Policy Makers

Organizations and think tanks are active in the sociopolitical issue of climate change, with various points of view and suggested policy action. Most of this is centered in Washington DC, aimed at Congress. Here is a selection of some of the voices and their positions.

NATIONAL ASSOCIATION OF MANUFACTURERS

Actively engaged in congressional proposals related to climate change, energy and the environment, NAM put forward these guidelines in its advocacy on behalf of American manufacturers:

- Preempt all state climate change laws so there is one national standard
- Support economic growth and do no harm to U.S. economy
- Be equitable and economy-wide in scope and include all sectors
- Recognize the different competitive environments and abilities of sectors
- Employ the most cost-effective implementation mechanisms
- Promote advanced, energy efficient and zero-and-low-GHG emission and sequestration technologies as part of a long-term strategy
- Enable Congress to set realistic deadlines that will accommodate the commercial deployment of necessary technology
- Address all significant greenhouse gases, sources and sinks
- Provide a reasonable level of flexibility for implementation, compliance and deadlines to aid in transitioning
- Include a safety valve or other equally effective and responsive cost containment mechanism
- Address the legal, technical, regulatory, liability, infrastructure framework and other challenges for capture, transport and

storage of carbon dioxide
- Encourage the development, funding and deployment of energy-efficient, zero-and low-carbon emission and sequestration technologies
- Expand the production and use of reliable, affordable and diverse domestic energy supplies
- Provide for free-standing legislation independent of the Clean Air Act's administrative, procedural and other provisions
- Harmonize with other mandatory climate change policies to avoid duplicating and conflicting requirements
- Promote global participation and encourage comparable global emission reduction action by all U.S. trading partners in a reasonable timeframe
- Recognize that fossil fuels are sometimes used for feedstock purposes and should be treated accordingly
- Should not exacerbate high U.S. natural gas prices, which are a major reason for 3 million lost manufacturing jobs in recent years
- Encourage the use of voluntary emission reduction initiatives and ensure that companies will not be disadvantaged later for current voluntary actions
- Give consideration to industries exposed to foreign competition if a U.S. climate change policy creates competitive disadvantages
- Avoid multiple and extreme penalties that adversely impact economic sectors
- Address the use of credits from non-participating countries
- Ensure transparency and open communication of policy, costs, benefits and uncertainties

U.S. CHAMBER OF COMMERCE

In its representation of the business community the U.S. Chamber advances these principles:

- Address the international nature of global climate change
- Promote accelerated technology development and deployment
- Preserve American jobs and the economy

Appendix

- Reduce barriers to development of climate-friendly energy sources
- Promote energy efficiency measures

UNITED STATES CLIMATE ACTION PARTNERSHIP

This partnership of leading corporations and green organizations, organized in Washington, DC, in 2006, advocates congressional action on for mandatory measures to reduce greenhouse gases, favoring a cap and trade system that will provide some degree of certainty and an orderly emissions trading market. USCAP favors a greenhouse gas registry and emphasizes that any legislation should assure adequate time for research and development. As members of USCAP, Ford, General Motors, and Chrysler, along with other U.S. companies, have agreed to cut CO_2 emissions from their cars and light trucks by 60–80% by 2050.

THE HEARTLAND INSTITUTE

This free-enterprise think tank, headquartered in Chicago, urges legislators to:
- Oppose higher energy taxes or emission caps on the grounds that the hypothetical benefits of reducing greenhouse gas emissions are too small to justify the cost.
- Support programs aimed at improving our understanding of climate and how it responds to changes in emissions and land use, perhaps as a way to produce research independent from national research programs that are biased toward alarmism
- Remove barriers to energy conservation embedded in state and local laws and regulations, such as restrictive building codes and zoning ordinances.
- Support research and limited capital investments if appropriate in adapting to climate change rather than trying to prevent it. If rising sea levels are a concern, for example, focus attention on the quality and reliability of ongoing investment in dikes.
- Pursue win-win strategies that produce enough benefits to pay their way apart from their possible effect on climate. For example, conservation tillage reduces the need for fertilizer and irrigation and helps prevent soil erosion. Tree planting can create recreational benefits.

COMPETITIVE ENTERPRISE INSTITUTE

CEI stated has long counseled "that the U. S. enact only those policies to address global climate change which are valuable in their own right—such as eliminating subsidies that promote the excessive use of fossil fuels and encouraging increased efficiency in energy production through the introduction of competition in the electric utility industry." CEI maintains that wide-scale policies to forcibly reduce CO_2 would impose heavy costs, increase the price of energy and produce a regressive impact, stifling American economic productivity.

NATIONAL MINING ASSOCIATION

NMA (which includes coal mining) member companies pledge to conduct their activities in a manner that recognizes the needs of society and the needs for economic prosperity, national security and a healthy environment.

Accordingly, we are committed to integrating social, environmental, and economic principles in our mining operations from exploration through development, operation, reclamation, closure and post closure activities, and in operations associated with preparing our products for further use.

This involves:
- Recognizing that the potential for climate change is a special concern of global scope that requires significant attention and a responsible approach cutting across all three of the sustainable development pillars: environmental, social and economic;
- Encouraging climate policies that promote fuel diversity, development of technology and long-term actions to address climate concerns in order to ensure that technological and financial resources are available to support the needs of the future; and,
- Supporting additional research to improve scientific understanding of the existence, causes and effects of climate change and to enhance our understanding of carbon absorbing sinks; advancements in technology to increase efficiencies in electrical generation and capture and sequester carbon dioxide;

Appendix

voluntary programs to improve efficiency and reduce greenhouse gas emission intensity; and, constructive participation in climate policy formulation on both international and national levels.

EDISON ELECTRIC INSTITUTE

Congress approved a far-reaching energy bill (H.R. 6) in mid-December 2007 and sent it to President Bush, who signed the legislation into law on December 19, 2007. The bill contains favorable energy-efficiency, smart grid, and carbon dioxide provisions, as well as incentives for plug-in hybrid electric vehicles. These provisions should be very helpful in the electric power industry's efforts to reduce greenhouse gas emissions.

CHAIRMAN DINGELL'S CARBON TAX PROPOSAL

U. S. House Energy and Commerce Committee Chairman John Dingell in 2007 provided details and intended impact of a carbon tax bill that he had drafted, and asked for public comment. He was at the same time engaged in more comprehensive, economy-wide approaches to climate change and energy matters. His leadership influenced congressional enactment of the 2007 energy bill with requirements for car and truck fuel efficiency. Following are excerpts from Rep. Dingell's Web site on the carbon tax proposal, in September 2007.

The Earth is getting warmer and human activities are a large part of the cause. We need to act in order to prevent a serious problem. The world's best scientists agree we need to reduce greenhouse gas emissions by 60–80 percent by 2050 in order to limit the effects of global warming and this legislation will put us on track to do just that. This is a massive undertaking, and it will not be easy to achieve, but we simply must accomplish this goal; our future and our children's futures depend on it. In order to get to this end we need to have a multi-pronged approach. In addition to an economy wide cap-and-trade program, which would mandate a cap on carbon emissions, a fee on carbon is the most effective way to curb emissions and make alternatives economically viable.

The legislation I am proposing would impose the following:
- A tax on carbon content: $50/ton of carbon (phased in over 5 years and then adjusted for inflation). Covering coal, including lignite and peat, petroleum and any petroleum product, natural gas.
- A tax on gasoline: $0.50/ gallon of gas, jet fuel, kerosene (petroleum based), etc. (added to current gas tax) (phased in over 5 years and then adjusted for inflation) The $0.50 gas tax is in addition to what is derived from the per ton carbon tax in the previous bullet.
- Exemption for diesel. The fuel economy benefits of diesel surpass even its emissions benefits; it provides about a thirty percent increase in fuel economy and a twenty percent emissions reduction.
- Biofuels that do not contain petroleum are exempt. Biofuels blended with petroleum are only taxed on the petroleum portion of the fuel.
- Phase out the mortgage interest deduction on large homes. These homes have contributed to increased sprawl and longer commutes. Despite new homes in and of themselves being more energy efficient, the sheer size, sprawl and commutes lead to dramatically more energy use – or to put it more simply, a larger carbon footprint. Specifically, the proposal:
- Phases out the mortgage interest on primary mortgages on houses over 3000 square feet.
- Exemptions for historical homes (prior to 1900) and farm houses.
- Exemptions for home owners who purchase carbon offsets to make home carbon neutral or own homes that are certified carbon neutral.
- An owner would receive 85% of the mortgage interest deduction for homes 3000–3199 square feet
- 70% for homes 3200–3399 square feet
- 55% for homes 3400–3599 square feet
- 40% for homes 3600–3799 square feet
- 25 % for homes 3800–3999 square feet
- 10% for homes 4000–4199 square feet
- 0 for homes 4200 square feet and up Where will the revenue go?

Appendix

- First and foremost, the Earned Income Tax Credit will be expanded. This helps lower income families compensate for the increased taxes on fuels.
- Expansion of the Earned Income Tax Credit
- Zero Children: Max earned income level from $5,590 to $7000; phase-out from $7000 to $9000
- One Child: Max earned income level from $8390 to $10,000; phase-out from $15,390 to $17,000
- Two or More: Max earned income level from $11,790 to $15,000; phase-out from $15,390 to $18,000
- The revenue from the gas tax goes into the highway trust fund, with 40% going to the mass transit and 60% going to roads. The revenue from the tax on jet fuel goes into the airport and airway trust fund.
- Finally, the revenue from the fee on carbon emissions will go into the following accounts: Medicare and Social Security; Universal Healthcare (upon passage); State Children's Health Insurance Program; Conservation; Renewable Energy Research and Development; Low Income Home Energy Assistance Program.

TERRA CHOICE'S "SIX SINS OF GREENWASHING"[124]

A 2007 report by TerraChoice Environmental Marketing generated considerable publicity by claiming that an overwhelming majority of environmental marketing claims in North America are "inaccurate, inappropriate, or unsubstantiated." Using metrics from the Federal Trade Commission (FTC) and the Environmental Protection Agency (EPA), TerraChoice concluded that all but one of the claims, out of more than 1000 products reviewed, raised red flags. Here are TerraChoice's "six sins of greenwashing" that got media and Internet attention:

Sin of the Hidden Trade-Off Example: Paper (including household tissue, paper towel and copy paper): "Okay, this product comes from a sustainably harvested forest, but what are the impacts of its milling and transportation? Is the manufacturer also trying to reduce

---

[124]See http://www.terrachoice.com.

those impacts?" Emphasizing one environmental issue isn't a problem (indeed, it often makes for better communications). The problem arises when hiding a trade-off between environmental issues.

Sin of No Proof Example: Personal care products (such as shampoos and conditioners) that claim not to have been tested on animals, but offer no evidence or certification of this claim. Company web sites, third-party certifiers, and toll-free phone numbers are easy and effective means of delivering proof.

Sin of Vagueness Example: Garden insecticides promoted as "chemical-free." In fact, nothing is free of chemicals. Water is a chemical. All plants, animals, and humans are made of chemicals, as are all of our products. If the marketing claim doesn't explain itself ("here's what we mean by 'eco' ..."), the claim is vague and meaningless. Similarly, watch for other popular vague green terms: "non-toxic," "all-natural," "environmentally-friendly," and "earth-friendly."

Sin of Irrelevance Example: CFC-free oven cleaners, CFC-free shaving gels, CFC-free window cleaners, CFC-free disinfectants. Could all of the other products in this category make the same claim? The most common example is easy to detect: Don't be impressed by CFC-free! Ask if the claim is important and relevant to the product. Comparison-shop and ask competitive vendors.

Sin of Fibbing Example: Shampoos that claim to be "certified organic," but for which our research could find no such certification. Is the claim true? The most frequent examples in this study were false uses of third-party certifications. Legitimate third-party certifiers—EcoLogoCM, Chlorine Free Products Association (CFPA), Forest Stewardship Council (FSC), Green Guard, and Green Seal, for example,—all maintain publicly available lists of certified products. Some maintain fraud advisories for products that are falsely claiming certification.

Sin of the Lesser of Two Evils Example: Organic tobacco. "Green" insecticides and herbicides.

Is the claim trying to make consumers feel "green" about a product category that is of questionable environmental benefit? Consumers

Appendix

concerned about the pollution associated with cigarettes would be better served by quitting smoking than by buying organic cigarettes. Consumers concerned about the human health and environmental risks of excessive use of lawn chemicals might create a bigger environmental benefit by reducing their use than by looking for greener alternatives.

The Page Principles[125]

> The Arthur W. Page Society and its senior communications executives are natural references for me in this book, since I have been a member of the group (incorporated in 1983) since 1989 and have served as officer, trustee and executive director. Our patron Arthur W. Page served as vice president of public relations for the American Telephone and Telegraph Company from 1927 to 1946, the first public relations executive to be an officer and board member of a major public corporation. We credit Page as having laid the foundation for securing corporate public relations as a top management function. In the Page Society's programs—such as the study of CEOs and corporate authenticity referred to in this book— the following seven principles of public relations management, drawn from Page's work and philosophy, are continually advanced:

- Tell the truth. Let the public know what's happening and provide an accurate picture of the company's character, ideals and practices.
- Prove it with action. Public perception of an organization is determined 90 percent by what it does and 10 percent by what it says.
- Listen to the customer. To serve the company well, understand what the public wants and needs. Keep top decision makers and other employees informed about public reaction to company products, policies and practices.
- Manage for tomorrow. Anticipate public reaction and eliminate practices that create difficulties. Generate goodwill.

---

[125]For more information on Arthur Page and the Arthur W. Page Society, see http://www.awpagesociety.com.

- Conduct public relations as if the whole company depends on it. Corporate relations is a management function. No corporate strategy should be implemented without considering its impact on the public. The public relations professional is a policymaker capable of handling a wide range of corporate communications activities.
- Realize a company's true character is expressed by its people. The strongest opinions—good or bad—about a company are shaped by the words and deeds of its employees. As a result, every employee—active or retired—is involved with public relations. It is the responsibility of corporate communications to support each employee's capability and desire to be an honest, knowledgeable ambassador to customers, friends, shareowners and public officials.
- Remain calm, patient and good-humored. Lay the groundwork for public relations miracles with consistent and reasoned attention to information and contacts. This may be difficult with today's contentious 24-hour news cycles and endless number of watchdog organizations. But when a crisis arises, remember, cool heads communicate best.

Index

Abbott Laboratories, 132
Abu Dhabi, 105-06
Abu Dhabi Future Energy Company, 105
Abzug, Bella, 40
Accountability
    of corporate executives, 56
    sustainability and, 51-55
ACE Insurance, 88
Advanced Micro Devices, 180
Advertising, public relations and, 139-40, 154-55
Advertising Council, 153, 154
AFL-CIO, 84, 92
Allegheny Energy, 92
Alliance for Climate Protection, 81
American Airlines, 91
American Conservative Union, 93
American Electric Power (AEP), 21, 113-14, 180-1
Anderson, Ray, 50, 70
Aqua International, 66
Arthur W. Page Society, 15, 143, 146, 158n, 158, 159, 235-36
    Page Principles, 235-36
ASTM International, 141
AT&T, 90, 167, 181
Automotive industry, CAFE regulations and, 119
Ayers, Dick, 64n

Baker, James A., 79
Bank of America, 91, 98
Bank of Montreal, 21
Baxter International, 109
Bayer, 109
Bed Bath & Beyond, 88
Beer, formaldehyde additives, 149
Ben & Jerry's, 90
Blood, David, 82
Blue Source, 114
The Body Shop, 90
Bolsinger, Lorraine, 195
Boxer, Barbara, 87
BP (formerly British Petroleum), 85, 131, 147, 155, 181-83
Brady, Todd, 201
Brazil, rainforest protection, 112
Brin, Sergey, 173
Brown, Jerry, 40
Burson, Harold, 97
Bush, George H. W., 34, 39, 66
Bush, George W., 39, 40
Business Council for Sustainable Development, 140
Business for Social Responsibility, 69
Business Roundtable, 21, 196
    RESOLVE program, 132, 196

Cadbury Schweppes, 103, 235
CAFE (corporate average fuel economy) regulations, 119
California, 207
    automotive efficiency regulatory authority, 44
    Oakland, 109
California Public Employees' Retirement System, 84
California State Teachers' Retirement System, 89, 170
Calvert Group, 88, 92
Calvert Social Research, 126
Canada, 116
    British Columbia, 116
    carbon taxes, 116
    Quebec, 116
Carbon dioxide (CO2), 49, 227
Carbon Disclosure Project (CDP), 20, 83, 86, 103, 122, 125, 135, 140, 171, 185, 234-35
    Supply Chain Leadership Collaboration, 103, 235
Carbon emissions, cap-and-trade schemes, 108, 114-15, 121
Carbon emissions offsets, 111-12, 114-15
Carbon emissions trading, 109, 114
    Chicago Climate Exchange, 68, 109-10, 125-26, 203, 217
    in Europe, 108-09
    regional, 110-11
Carbon neutrality, 112-13
Carbon taxes, 115-16, 117-19, 229-31
    in Canada, 116-17
Carbon war, 11-12, 18, 172, 173
    company rankings, 135
    strategies, 129-34
    tracking mechanisms, 41
Carbonomics, 10, 105-06
    CCO role in, 107
    defined, 38
    in Europe, 108-09
    in the U.S., 106-08
CarbonTracker system, 41, 228, 229
Carpenter, George, v, 208
Carson, Rachel, 2, 29, 32, 44, 106, 166
Carter, Jimmy, 29, 30, 32
Caterpillar (CAT), 93-94, 115, 130, 156, 157, 170, 183
Catholic Healthcare West, 91
CEQ. See President's Council on Environmental Quality
Ceres (formerly Coalition for Environmentally Responsible Economics), 19, 20, 27, 88, 91, 94, 125
    See also Valdez Principles
Charles (Prince of Wales), 22

Chevron, 98, 116
Chicago Climate Exchange (CCX), 109-10, 125-26, 203, 217
Chilcott, Lesley, 112
China, 149, 151
　carbon dioxide emissions, 11
　greenwashing and, 149-50
Chlorofluorocarbons (CFCs), 154, 228, 233
Cinergy Corporation, 21
Cisco Systems, 98
Cities, Yahoo's 2007 greenest list, 57, 233-34
Citigroup, 86, 98, 116, 132, 183
Clean Air Act, 64, 65
Clean Air-Cool Planet, 135
Clean Energy Group, 206
Climate change issues
　advice to policy makers, 223-36
　corporate positions on, 179-219
　partnerships with government, 68
　partnerships with NGOs and, 8-9
　public expectations on, 157-58
　scientific vs. sociopolitical aspects, 3-4
　tips for corporations, 15, 176
Climate Counts, 135, 142, 170
Climate Savers Computing Initiative, 8, 200
Climatewire (Internet news service), 85
Clinton, Bill, 117
Clinton Climate Initiative, 218n, 218
Coca-Cola Company, 183-85
Cody, Iron Eyes, 153
Collaboration, 66-67
　with government, 20-27, 69, 132, 142
　IBM as model for, 68
　with NGOs, 8-9, 53, 143
Collins, Jim, 6, 7
Combat Climate Change (3C), 206
Community investing, 102
Competitive Enterprise Institute, 93, 226
Cone, Carol, 75n, 75
Connecticut, State Treasurer's Office, 88
ConocoPhillips, 87, 92, 156
The Conservation Foundation (TCF), 65-66
CONSOL Energy, 92
Consumers, expectations from industries, 124
Corporate communications, 23-25, 158-61
　Academy Awards lessons for, 77-80
　activists' hostility, 137-38, 148
　advertising vs. public relations, 139-40, 154-55
　backlash, 153-55
　on climate change issues, 123-28, 131-32, 151
　creating stakeholders, 127
　Earth Day and, 169
　evironmental activism and, 92-93
　on green strategies, 132-34, 163
　internal, 97-98
　public expectations and, 157-58, 161
　with stakeholders, 6
　on sustainability, history, 35-37
　use of World Wide Web for, 127-28
　See also Green strategies; Greenwashing; Transparency
Corporate communications officers (CCOs), 12-13, 55-56, 119-20, 161
　carbonomics and, 107

QUALITY model, 72-73
　role of, 6-7, 60
　strategy questions for, 60-62
　sustainability challenges, 71
Corporate greening, 4-5, 17
　executives role in, 20-23
　See also Green strategies
Corporate Greening 2.0, 13-14, 73, 108
　Greening 1.0 compared, 12

Corporate social responsibility (CSR), 24
Corporate sustainability. See Sustainability
Cousteau, Jacques, 40
Covey, Stephen R., 156
CSR. See Corporate social responsibility

DaimlerChrysler, 185
Dalai Lama, 40
Daly, Patricia A., 95
Darbee, Peter A., 16, 206
David, George, 42
Dell, 8, 103, 132, 185-86, 200, 235
　Zero Carbon Initiative, 185
Denver, John, 40
DiCaprio, Leo, 77
Dingell, John D., 16, 22, 117, 118-19, 229
Domini 400 Social Index, 101
Dominion Resources, 88
Dow Chemical Company, 90, 131, 186
Dow Jones, 98
　STOXX Global 1800 Index, 124
　Sustainability Index (DJSI), 98, 99, 126, 191
Drucker, Peter F., 62, 152, 160, 162
Dubai Cargo, Village Environmental Awareness Award, 192
Duke Energy Corporation, x, 65, 115, 130, 131, 156, 187-88
DuPont Company, 65, 85, 90, 130, 139, 154-55, 156

E&ETV (Internet news service), 85
Earth: The Sequel (Krupp), 64
Earth Day, 45, 167
　origins, 165-66
　(1970), 44, 165
　(2007), 38
　(2008), 167-68
Earth Day Network, 168
Earth Summit (Rio de Janeiro 1992), 2, 4, 33-34, 45, 90, 124, 138, 140, 144, 150
Eaton Corporation, 192
Eckhart, Michael, 152, 159
The Economist (magazine), 40, 104, 173
Edison Electric Institute (EEI), 22, 227
Electronic Data Systems (EDS), 8, 200
ELFA system, 213
Emissions trading. See Carbon emissions trading; Greenhouse gases
Energy
　alternative sources, 10
　nonrenewable resources, 10
Energy efficiency, profit in, 27, 95, 129, 231
Energy Future Holdings, 66n, 66
　See also TXU Energy
Energy Policy Act of 1992, 126

ENERGY STAR program, 126
Entergy, 156, 191-93
EnviroComm International, 2, 131
   QUALITY model, 72
Environmental activism, 18-19, 166
   corporate communications and, 92-93
   corporate confrontations with, 85-89
   history, 29-34, 43-44, 73
   hostility in, 137-38, 148
   by investors, 19-20
   opinion polls and, 37-38
   See also Sustainability
Environmental Credit Corporation, 113
Environmental Defense Fund (EDF), 8, 63, 143, 144, 192
EPA. See U.S. Environmental Protection Agency
Esty, Daniel C., 47
Ethanol, 9-10
Europe, carbon trading in, 108-09
European Union (EU), cap-and-trade scheme, 108, 121
Exchange-traded funds (ETFs), 102
Exxon Valdez, 19, 89
ExxonMobil Corporation, 87, 92, 158, 170, 189-90

Federal Trade Commission, Environmental Marketing Claims Guidelines, 141, 146
FedEx Corporation, 116, 145, 192-93
Fisher, Roger, 72
Florida Power and Light (FPL) Group, 193
Fluorinated gases, 49, 227
Fonda, Jane, 40
Forbes (magazine), 191
Ford Motor Company, 22, 46, 89, 91, 94-95, 109, 156
   climate change strategy, 193
Fortune Brands, 193-94
Fortune (magazine), 43, 99
Frankle, Janice Podoll, 141
Free Enterprise Action Fund, 94
Freeport-McMoRan, 2
Friedman, Thomas, 47
Friends of the Earth, 93
Fukuyama, Francis, 60
FutureGen, 10

G-7 Group, 39
G-8 Group, 40
Gap, 125
Gases. See Greenhouse gases; names of specific gases
Gates, Bill, 174
General Electric Company (GE), 10, 47, 98, 115, 125, 130, 131, 156, 158
   Citizenship Report, 123
   Ecomagination program, 147, 195
   "Ecomagination" commercial, 139-40
General Mills, 196
General Motors (GM), 1-2, 52, 88, 91, 126, 128, 130, 145, 156, 196-97
George Mason University, 38
Getting to Yes (Fisher & Ury), 72
Ghostbusters (film), 96, 97
Gingrich, Newt, 31n, 31, 32
Global Environment Management Initiative (GEMI), 192
Global Reporting Initiative (GRI), 125, 188
   Sustainability Reporting Guidelines, 125
Going green, terminology, 45, 47-48
Going Green (Harrison), 2, 46, 155, 162
Goldman Sachs, 20, 157
Golightly, Neil, 193
Google, 8, 200
Google.org, 173
Gore, Al, 1, 3n, 36, 37, 40, 73, 77-78, 82, 144, 169
   An Inconvenient Truth film, 112, 113
   "We" campaign, 81
Government
   business partnerships with, 20-27, 69, 132, 142
   See also names of specific government agencies
Green Mountain Power, 109
Green movement, sociopolitical aspects, 29
Green strategies
   carbon war, 129-34
   of specific companies, 179-219
   See also Corporate communications; Corporate Greening; Sustainability
Greenberg, Jack, 63
GreenBiz Index, 143
Greenhouse Crisis Foundation, 79
Greenhouse Gas Protocol, 197
Greenhouse gases (GHGs), 49, 227-28
   emissions trading, 107-08
Greenpeace, 143, 146, 184
Greenwashing, 138-43, 145-46, 147, 148
   China and, 149-50
   six sins of, 232-33
   See also Corporate communications
Grist.org (Web site), 118, 135, 164

Halliday, Chad, 188
Harvard University, 192
Hawken, Paul, 96
Hawkins, David, 58, 63, 64-65, 79
Hayes, Denis, 166, 168
The Heartland Institute, 3n, 225
Hewlett-Packard Company (HP), 8, 48, 197-98, 200
High Global Warming Potential gases, 49
Hill, Gladwin, 42
Holtz, Robert Lee, 41, 228
Home Depot, 198
Honeywell International, 198-99
Hurricane Katrina, 12, 37, 144, 191

Illinois, Chicago, 109
Immelt, Jeffrey, 140, 170
An Inconvenient Truth (film), 112, 113
Institute of Management and Administration (IOMA), 20n
Institutional investors, 102
Intel Corporation, 8, 200-201
Interface, 70n
Interfaith Center on Corporate Responsibility, 88, 92
Intergovernmental Panel on Climate Change, 26
International Business Machines (IBM), 8, 94, 116, 123, 125, 126, 199-200
   as model collaborator, 68
International Emissions Trading Association, 141
International Standards Organization (ISO), 141,

International Truck and Engine, 53, 156
Internet. See World Wide Web
Investors
   environmental activism by, 19-20, 83-84, 88-89, 170
   information-gathering by, 97-98
   shareholder resolutions, 86-88
   social investment trends, 102
   sustainability actions and profit, 95
Iowa, 109
Iowa State University, 87

Jagger, Bianca, 40
Jenkins, Holman W., Jr., 115
John Deere & Company, 201-02
Johnson & Johnson, 98, 126, 156, 202-03
Johnson, Jack, 48
Johnson, Keith, 84
JPMorgan Chase, 86, 114

Kaplan, David, 178
Kerr-McGee Corporation, 91
Kerry, John, 79
Khosla, Vinod, 104, 173
Kimberly-Clark, 146
Kinder, Peter D., 101
KLD Research & Analytics, 101
KPMG, 124
Kravis, Henry, 79
Krupp, Fred, 58, 63-64, 171
Kyoto Protocol (1992), 108

Lane, Robert W., 201, 202
Lash, Jonathan, 59, 156, 163, 178
Lash-Wellington climate change guide for business, 163
Laurie, Bob, 166
Laurie, Marilyn, 15, 104, 165, 166, 167, 169, 176
Leadership in Energy and Environmental Design (LEED), Green Building Rating System, 141
Lenovo Group, 8, 200
Leonard, J. Wayne, 191
Lindsay, John, 165, 166
Longhurst, Mike, 136
Love Canal, 12
Lovins, Amory, 11
Lowery, Beth, 197
Lubber, Mindy S., 27, 92, 95, 129, 232
Lundberg, Kory, 178

Mackey, John, 55
Makower, Joel, 38, 135, 143, 148
Marland, Gregg, 113
Marriott International, 112
Massachusetts Institute of Technology (MIT), 105, 116
   Joint Program on Science and Policy of Global Change, 190
McCain, John, 169
McDonald's, 63, 144, 184
McKinsey & Company
   2007 survey, 17, 124
   energy efficiency investment study (2008), 27, 231

Meezan, Erin, 70
Merrill Lynch, 20, 86, 171
Metcalf, Robert, 173
Methane (CH4), 49, 227
Microsoft Corporation, 8, 98, 174, 200
Miller, John, 229
Mission statements, green, 90-91
Molawad, Jad, 28
Monsanto Corporation, 203
Montreal Protocol on Substances that Deplete the Ozone Layer (1987), 106, 139, 154
Moore, Roger, 40
Morgan Stanley, 86
Motorola, 109
Murdoch, Rupert, 205
Murray Energy Corporation, 93
Musk, Elon, 173

Napolitano, Janet, 78
National Association of Manufacturers, 21, 223-24
National Center for Public Policy Research (NCPPR), 93
National Mining Association, 226-27
National Oceanic and Atmospheric Administration (NOAA), 229
   CarbonTracker system, 41, 228, 229
National Wildlife Federation, 8, 93
Native Energy, 115
Natural Resources Defense Council, 8, 64, 77, 79
Navistar International Corporation, 203-04
   International Truck and Engine, 53, 156
Nebraska, 109
Nelson, Gaylord, 42, 165, 168
Nestlé, 103, 204-05, 235
New Mexico, 109
New York City
   Comptroller's Office, 88
   Mayor's Council on the Environment, 167
New York State Common Retirement Fund, 122
New York Times (newspaper), 52, 85, 155, 167
Newman, Paul, 165, 167
News Corporation, 205
Nike, 125
Nitrogen oxide (NO2), emissions trading, 108
Nitrous oxide (N2O), 49, 227
Nixon, Richard, 32
Nongovernment organizations (NGOs), corporate alignments with, 8-9, 53, 143
Nordhaus, Ted, 28, 31
Northeast Utilities, 91
Novelli, Porter, 96
Novo Nordisk, 158

Obama, Barack, 169
O'Donnell, Frank, 64n
Ohio, Toledo, 9
Opinion surveys, effects of, 37-38
The Oprah Winfrey Show, 43, 46
Oregon, Portland, 109
Owens, James, 93, 94

Pacific Gas and Electric Corporation (PG&E), 131, 200, 205-07

Page, Arthur W., 5, 13, 47n, 47, 53, 150, 161, 235
Page, Larry, 173
Page Principles, 235-36
Partnerships. See Collaboration
Patagonia, 158
People's Republic of China. See China
Perot Systems, 6
Petroleum, 10
Pfizer, 98, 132, 207
Pizer, William A., 109
Policy makers, climate change advice to, 223-36
Pope, Carl, 31, 37
Powell, Colin, 5, 6
President's Council on Environmental Quality (CEQ), 65
Prince of Wales's Business and the Environment Programme, 50n
PRNews, 122
Procter & Gamble, 90, 98, 208
Prudential, 208
Public Affairs Council, 29, 30
Public Environmental Reporting Initiative (PERI), 123
Public opinion. See Surveys (opinion)
Public relations, 172
    advertising and, 139-40, 154-55
Public Relations Seminar, 15
Public transportation, shift to, 11

QUALITY model, 72

Radio talk shows, 85
Rain Forest Action Network, 145
Raytheon, 208-09
Regional Greenhouse Gas Initiative (RGGI), 110-11
Reilly, William K., 40, 65, 79
Renewable energy offsets, 113-14
Resources for the Future (RFF), 117
Responsible Care, 52, 137, 143
Ridenour, David, 93, 94
Ries, Al, 52
Rifkin, Jeremy, 79, 80
Rio Tinto, 210
Rockefeller family, 87, 170
Rocky Mountain Institute, 109
Rosenzweig, Phil, 23
Rossen, Matt, 57, 234
Royal Dutch Shell, 90, 131, 156, 158, 211-12

SAM Group Holding, sustainability research group, 100, 126
Sarbanes-Oxley regulations, 56, 98
Sauers, Len, 70
Schwarzenegger, Arnold, 73, 78-79
Schwarzman, Stephen, 79
Scott, Lee, 50, 218, 219
Secrecy, in C-Suites, 55-56
Seeger, Pete, 167
Service Employees International Union (SEIU), 87
Shabecoff, Phil, 36, 153
Shareholder resolutions, 102, 170
Shell. See Royal Dutch Shell
Shellenberger, Michael, 28, 31
Siemens Energy & Automation, 212-13

Sierra Club, 30, 31
Sierra Club Mutual Fund, 88
Sierra Student Coalition, 30
Silent Spring (Carson), 2, 29, 35, 106, 166
Sisters of St. Dominic, 95
Sloan, Jim, 122, 127
Smith, Cheryl, 101
Social Investment Forum (SIF), 86, 100-101
Socially responsible investing (SRI), 100-102
Society of Environmental Journalists, 30, 137
Sony Electronics, 213
Southern Company, 21
Stakeholders, 3, 5-6, 7-8, 79, 127
    belief in a company's green commitments, 75-76
    communications with, 51-52
    executive mindset and, 62
    perceptions of, 161-62
Standards
    companies' green operating standards, 141-42
    environmental, 52
    ISO 14000, 52, 141
Stanford University, 38
    Global Climate and Energy Project, 189, 190
State of Green Business 2008 (Makower et al.), 143
Stonyfield Farms, 135
Sulfur dioxide (SO_2), emissions trading, 107
Sun Microsystems, 173
Surveys (opinion), 37-38
    by ABC News, 38
    on belief in corporate green commitments, 75
    by George Mason University (2008), 38
    McKinsey survey (2007), 17
    SAM survey of corporate sustainability, 100
    by Stanford University, 38
Sustainability, 4, 7, 9-10, 23-25, 45, 75-76, 169-72
    accountability and, 51-55
    business uncertainties and, 54
    companies' positions on, 179-219
    corporate philanthropy and, 8
    Dow Jones rating of, 98-99
    financial performance and, 54-55
    fiscal responsibility and, 116
    global business implications, 11
    green mission statements, 90-91
    investors' focus on, 19-20, 83-84
    involving stakeholders in, 79
    mainstreaming of, 73
    officer, 46
    plan guideposts, 74-75
    profit and, 27, 95, 129, 231
    reasons for, 43-45
    socioeconomic factors, 45-46
    survey of practitioners, 157
    technology and, 173-75
    See also Environmental activism
Sustainability Reporting Guidelines, 125
Sustainable Asset Management, 59
Sweeney, John, 84
Synge, Peter, 156

Technology, 26
    sustainability and, 10, 173-75
TECO Energy, 109
Television shows, 43, 46, 85

TerraChoice Environmental Marketing, 145, 232
Texas, 79
Texas Pacific Group, 66
Three Mile Island, 12
Tillerson, Rex W., 118, 189-90
Time (magazine), 43
Toyota, 214-15
Train, Russell, 65
Transparency, 13, 140, 144-45
　C-suite secrecy and, 55-56
　inviting inspections, 52
　See also Corporate communications
Transportation, 10-11
　shift to public, 11
Tree planting, 111-12, 176
Tri-State Coalition for Responsible Investment, 95
Trillium Asset Management, 87, 92
Trout, Jack, 52
Tufts University, 109
Turner, Ted, 40
TXU Energy, 8, 10, 21, 63, 66n, 66, 79-80, 92, 144
　See also Energy Future Holdings
Tyco International, 21

Unilever, 103, 184, 235
Union of Concerned Scientists, 93
United Arab Emirates, 105
United Kingdom, Advertising Standards Authority, 146
U.N. Conference on business impacts of government decisions (New York, February 2008), 84, 95
U.N. Conference on Environment and Development (UNCED; Rio de Janeiro 1992). See Earth Summit
U.N. Environment Programme, 125
U.N. Environment Programme (UNEP), 184
United Parcel Service (UPS), 216-17
U.S. Chamber of Commerce, 21, 224-25
U.S. Climate Action Partnership (USCAP), 8, 21, 22, 63, 64, 87, 93, 114, 131, 145, 148, 183, 206, 225
U.S. Climate Change Science Program, State of the Carbon Cycle Report, 41, 228
U.S. Congressional Budget Office, 93
U.S. Department of Defense, 142
U.S. Department of Energy, 116, 142, 228
　Voluntary Greenhouse Gas Emissions Reporting, 68, 125, 126
U.S. Environmental Protection Agency (EPA), 32, 44, 45, 64, 90, 114, 135, 142, 146, 200
　Climate Leaders program, 20-27, 132, 142
　Climate Partnership Programs, 69
　Climate Wise Buildings program, 126
　ENERGY STAR program, 126, 209
　Green Power Leadership Award, 192
　Green Power Partner program, 200

National Greenhouse Gas Inventory, 41
National Greenhouse Gase Inventory, 228
National Partnership for Environmental Priorities, 142
Performance Track, 142
　sulfur dioxide emissions and, 107
U.S. Overseas Private Investment Corporation, 66
United Technologies (UTC), 132, 217-18
University of California Los Angeles (UCLA), 116
University of Minnesota, 109
UOP/Eni Ecofining process, 199
Ury, Bill, 72

Valdez Principles, 89-90, 91-92, 125
　See also Ceres
van der Veer, Jeroen, x, 212
Viera, John, 82
Village Voice (newspaper), 166

Wagoner, Richard, 1, 3
Wal-Mart, 8, 54, 63, 144, 156, 157, 170, 178, 218, 219
　Sustainability 360, 55
Wall Street Journal (newspaper), 115, 228
Walsh, Brian, 43
War on carbon. See Carbon war
"We" campaign, 81
Web sites, 135, 142
Wellington, Fred, 156, 163, 178
Wells Fargo, 87-88, 92
Wenzel, Mary S., 92
Werbach, Adam, 30-31, 32
WestStart-CALSTART, 192
Winston, Andrew S., 47
World Business Council on Sustainable Development, 34, 71, 90, 141, 197
World Economic Forum, Global Greenhouse Gas Register, 185, 197
World Resources Institute, 59, 109, 197
World Wide Web, business and climate change coverage on, 85, 86
World Wildlife Fund, 8, 66, 197, 200

Yahoo, greenest cities list, 57, 233-34
Yarnold, David, 136, 145, 148
Year 2000 computer challenge (Y2K), 26
Yosie, Terry, 123

Zoi, Cathy, 164
Zucker, David, 96

About the Author

Bruce Harrison has counseled senior management in more than fifty corporations during his career as corporate vice president in New York, counseling firm owner in Washington, DC, and franchiser of the EnviroComm International network in Europe.

Bruce began his career as newspaper reporter in Georgia and press secretary to a member of Congress in Washington and Alabama. He entered public affairs in the private sector in 1962 by joining the chemical industry association, the year that *Silent Spring* by Rachel Carson launched America's environmentalism era, where he was appointed vice president and environmental information officer—the first time such a title had appeared in the business community. Bruce subsequently joined Freeport Minerals Company (now Freeport-McMoran) in New York, as vice president and chief communications officer. In addition to international public relations, which included the launch of a copper mining project in Indonesia, his responsibilities were expanded to cover investor relations and government relations.

Returning to Washington and entering the private sector, Bruce founded the EnviroComm counseling practice. His book, *Going Green*, published in 1992, provided early guidance on sustainable business communications. He was part of the business delegation at the first United Nations Earth Summit that year. In 2001, *Public Relations Week* recognized Bruce as one of the 100 most influential public relations people of the 20th century, for his work in environmental and social issues in the business community.

He may be accessed by email through http://www.envirocomm.com.